Lecture Notes in Physics

Edited by H. Araki, Kyoto, J. Ehlers, München, K. Hepp, Zürich
R. Kippenhahn, München, H. A. Weidenmüller, Heidelberg
J. Wess, Karlsruhe and J. Zittartz, Köln
Managing Editor: W. Beiglböck

281

Ph. Blanchard
Ph. Combe
W. Zheng

Mathematical and Physical Aspects of Stochastic Mechanics

Springer-Verlag Berlin Heidelberg GmbH

Authors

Ph. Blanchard
Theoretische Physik and BiBoS, Universität Bielefeld
D-4800 Bielefeld, FRG

Ph. Combe
BiBoS, D-4800 Bielefeld, FRG
and Université Aix-Marseille II and Centre de Physique Théorique
C.N.R.S. – Luminy-Case 907
F-13288 Marseille, France

W. Zheng
BiBoS, D-4800 Bielefeld, FRG
and East Normal University, Shanghai, China

ISBN 978-3-662-13606-5 ISBN 978-3-540-47718-1 (eBook)
DOI 10.1007/978-3-540-47718-1

Originally published by Springer-Verlag Berlin Heidelberg New York in 1987
Softcover reprint of the hardcover 1st edition 1987

P R E F A C E

These are the revised Lecture Notes of a graduate course given at Bielefeld University in the "Sommersemester" 1985. The purpose is to present a self-contained (apart from the inevitable basic prerequisites) version of the mathematical and physical aspects of stochastic mechanics. The course was very lively and we enjoyed the active collaboration of all participants, who contributed substantially to the book.

The Research Centre BiBos "Bielefeld-Bochum-Stochastics" of the Stiftung Volkswagenwerk should be especially thanked for sharing the costs and for coordinating the invitation of two of the lecturers.

The material of the lectures has been thoroughly reorganized and rewritten and we have benefited from close interaction with a number of friends and colleagues. We wish to record our indebtedness to S. Albeverio, G. Bolz, E. Carlen, O. Cohendet, G.F. Dell'Antonio, D. Dürr, M. Fukushima, D. Gandolfo, S. Golin, F. Guerra, A. Hilbert, R. Høegh-Krohn, G. Jona-Lasinio, R. Jost, R. Marra, M. Mebkhout, P.A. Meyer, H. Nagasawa, E. Nelson, H. Nencka-Ficek, J. Potthoff, R. Rodriguez, H. Rost, W. Schneider, M. Serva, M. Sirugue, M. Sirugue-Collin, O. Steinmann, L. Streit, J. Stubbe, J.C. Zambrini and M. Zhanghi for helpful advice and comments.

Special thanks go to secretaries, in particular to the secretaries at Bielefeld (Mrs. Jahns, Mrs. Jegerlehner, Mrs. Litchewski and Mrs. von Reder) for expert typing.

Bielefeld, April 1987

C O N T E N T S

VI. TWO VIEWPOINTS CONCERNING QUANTUM AND STOCHASTIC MECHANICS

VII. A NON-QUANTAL LOOK AT STOCHASTIC MECHANICS

APPENDIX

I. INTRODUCTION

I.1 Some Probabilistic Aspects in Classical and Quantum Physics

In physics probabilistic ideas and concepts occurred for the first time in connection with the statistical approach to thermodynamics, the so-called kinetic theory of gases, during the second part of the last century and in the publication of Albert Einstein's paper on Brownian motion in 1905 [42]. Einstein's theory not only provided a decisive breakthrough in the understanding of the phenomena of Brownian motion; in the opinion of Max Born it also did "more than any other work to convince physicists of the reality of atoms and molecules, of the kinetic theory of heat, and of the fundamental rôle of probability in the natural laws".

I.1a Brownian Motion. Mechanics of Particles Submitted to Random Disturbation

The motivation for Einstein's work on the theory of Brownian motion was "to find facts which would guarantee as much as possible the existence of atoms of definite finite size" [95]. The existence of atoms and molecules was postulated in the kinetic theory of gases some decades before by C. Maxwell, R. Clausius and L. Boltzmann.

In his "Autobiographical Notes" [95],Einstein indicated the relation of this work to the state of physics at the beginning of this century: "I discovered that, according to atomistic theory, there would have to be a movement of suspended microscopic particles open to observation, without knowing that observation concerning the Brownian were already long familiar". Indeed, this physical phenomenon was first described by R. Brown in 1827 [19].

Einstein proposed an experiment based on a theoretical model describing the erratic motion of a very small spherical particle through a viscous medium and submitted to the influence of thermal molecules of the bath, for which he assumed a molecular structure as in the kinetic theory. Since the frequency of the collision is very high ($10^{21}s^{-1}$), the velocity of the test particle changes very often and it is therefore impossible to measure the speed of particles submitted to Brownian motion. Taking into account that because of their size the velocity of the Brownian particles must be much smaller than those of the molecules, he concluded that the mean position of the Brownian particles is zero.

Hence Einstein proved that statistical fluctuations can produce sufficiently important effects to induce an erratic motion, which can be observed under a good microscope. From his model, Einstein concluded that after a sufficiently long time the random motion of the spherical particles does generate a migration. Moreover, he proved that this phenomenon is essentially a diffusion. The basis of his description is the notion that the suspended particles are "diffusing" through the liquid in such a way that the dynamical equilibrium is maintained between the osmotic force and the viscous force. For the probability density $\rho(x,t)$ of a Brownian particle to be in x at time t, Einstein obtains the equation

$$\frac{\partial \rho}{\partial t} = \nu \, \Delta \, \rho \tag{1.1}$$

where ν is a positive constant called the diffusion coefficient, which depends on the nature of the particle and of the properties of the liquid (viscosity, temperature, ...). This implies that the mean square displacement at time t is given by

$$\langle x^2 \rangle_t = 2\nu t \tag{1.2}$$

assuming that the particle starts from the origin at time $t=0$.

This result is strongly reminiscent of the random walk process: the root mean square distance travelled is proportional to $\sqrt{t}$.

Moreover, Einstein gives an explicit formula for the diffusion coefficient

$$\nu = \frac{kT}{m\beta} \tag{1.3}$$

where $m\beta$ is the resistance due to the friction, k the Boltzmann constant and T the absolute temperature.

The measurement of $\langle x^2 \rangle_t$ then yields a knowledge of $k = \frac{R}{N_0}$ where R is the constant of perfect gas, and by (1.3) one obtains Avogadro's number N_0. The first good determination of N_0 was made by J. Perrin in 1909 along this method [92].

The study of Brownian motion as a stochastic process was undertaken by N. Wiener in 1923 [110], preceded by L. Bachelier's heuristic work [10], and soon was developed into its modern form by Paul Levy and his followers, A. Kolmogorov [77], K. Ito [69], J.L. Doob [41]; K.L. Chung [25], M. Kac [75], P.A. Meyer [85a,b]. Together with the

Poisson process, it constitutes one of the two fundamental "species" of random processes, both in theory and applications.

Einstein's theory does not provide a dynamical theory of Brownian motion, this one appeared in a paper by M.V. Smoluchowski [101] on the same subject but following a completely different line of thought. Smoluchowski attempted a more dynamical theory.

If the force K acting on a particle is constant and denoting by $m\beta$ the friction coefficient, then (in classical mechanics) the particle should acquire the limit velocity K/β. After a long time (large compared with the relaxation time β^{-1}) the velocity will be approximately K/β. Taking now into account the random fluctuations due to the Brownian motion, this suggests the differential stochastic equation

$$dx(t) = \frac{K}{\beta}\,dt + \sqrt{2\nu}\,dW_t \tag{1.4}$$

where W_t is the standard Wiener process with covariance $\mathbb{1}$. This relation is the starting point of Smoluchowski's theory (expressed in modern terminology).

This approach to Brownian motion suggests naturally a model to describe the dynamics of a particle moving on one hand under the influence of external forces and on the other hand submitted to very rapidly varying forces due to the molecules that make up the medium in which the particle is immersed.

This point of view was proposed in 1911 by P. Langevin [78] to describe the erratic motion of a particle in a random environment. Applying the fundamental law of Newtonian mechanics to a particle of mass m submitted to viscous forces and random fluctuations, he writes

$$m\,\frac{d^2x}{dt^2} = m\beta v + mf,\quad v = \frac{dx}{dt} \tag{1.5}$$

where mf is a rapidly varying force, called "white noise" (this terminology is due to the fact that its spectrum contains all the frequencies with the same intensity). This white noise is the derivative in the weak sense of the Wiener process (the paths of Brownian motion being almost surely non-differentiable). Langevin's equation (1.5) connects therefore two worlds: the macroscopic world represented by the drag force and the microscopic world described by the fluctuating force.

Hence Langevin wrote what we call today a stochastic differential equation, getting

$$dx(t) = v(t)dt$$
$$dv(t) = -\beta v(t)dt + dW(t). \qquad (1.6)$$

This description of Brownian motion has been achieved by L.S. Ornstein and G.E. Uhlenbeck in the thirties [106], [108]. In some sense, this was the first step toward a mechanics of systems submitted to some random forces. Later developments led to stochastic mechanics introduced in 1966 by E. Nelson [90a] in connection with quantum mechanics. Nelson explained the dynamical signification of formal equations discovered by I. Fényes [45] using a stochastic version of Newton's equation. In fact, in this framework one is naturally led to the Schrödinger equation, starting from a description of microprocesses by means of diffusions.

I.1b From Feynman Path Integral to Probabilistic Formulations of Quantum Physics

The founding of quantum mechanics can be located between 1923 and 1927. The meaning of the wave function proposed by M. Born [18] (see also [71]) is that if a quantum mechanical system is described by the wave function ψ then $|\psi(x,t)|^2$ represents the probability density of finding the system in x at time t. Despite the similarity between Schrödinger's wave mechanics and diffusion theory, as we will see in §I2, it was clear since the very beginning of quantum mechanics [74] that within quantum theory a new kind of probability was involved.

In 1948, R.P. Feynman [47,48] proposed a global formulation of quantum mechanics based on probabilistic ideas, in which the propagators associated to time evolution are expressed as a functional integral, the so-called "Feynman path integral".

Let us sketch Feynman's idea briefly in the case of a particle under the influence of a potential $V(x)$. In classical mechanics the position of the particle is known at each time, and the particle describes a well-defined trajectory starting at time t_a from the point x_a and ending at time t_b on the point x_b. In quantum mechanics we introduce a probability amplitude $K(x_a,t_a,x_b,t_b)$. Feynman's idea was to write this amplitude as the sum of the contribution of all paths leaving x_a at time t_a and ending in x_b at time t_b ("sum over histories")

$$K(x_a,t_a;\, x_b,t_b) = \sum_{\substack{\gamma(t_a)=x_a \\ \gamma(t_b)=x_b}} \phi(\gamma). \qquad (1.7)$$

Here is the main difference to classical mechanics: for the classical particle there is only one possibility, the "classical path", for which the action functional

$$S(\gamma) = \int_{t_a}^{t_b} L(\gamma(t),\dot{\gamma}(t),t)\,dt \tag{1.8}$$

is stationary, L being the Lagrangian of the system.

Feynman's "Ansatz" consists in assuming that all paths γ such as $\gamma(t_a) = x_a$ and $\gamma(t_b) = x_b$ give a contribution with the same amplitude but with different phases. This phase is given by the action function along the path, i.e.,

$$\phi(\gamma) = Ae^{\frac{i}{\hbar}S(\gamma)} \tag{1.9}$$

where $\hbar$ is the Planck constant divided by 2π.

In Feynman's formulation of quantum mechanics we can therefore formally write

$$K(x_a,t_a,x_b,t_b) = N \int_{\substack{\gamma(t_a)=x_a \\ \gamma(t_b)=x_b}} e^{\frac{i}{\hbar}S(\gamma)} \mathcal{D}(\gamma) \tag{1.10}$$

where N should be a suitable normalization and $\mathcal{D}(\gamma)$ is the "measure", on the manifold of paths, which can be formally written as a product of Lebesgue's measure, e.g.

$$\mathcal{D}(\gamma) = \prod_{t\in[t_a,t_b]} d\gamma(t) \ . \tag{1.11}$$

The transition function K itself is not a probability but the square of its modulus defines a probability density

$$P(x_a,t_a,x_b,t_b) = |K(x_a,t_a;x_b,t_b)|^2 \ . \tag{1.12}$$

One of the interesting aspects of Feynman's formalism is the fact that it gives a completely independent and self-contained formulation of quantum mechanics and allows, at least in principle, direct extension of quantum field theory. Another appealing feature of this formalism is its strong connection with the Lagrangian formulation of classical mechanics, allowing a mathematical control on the approach to the classical limit $\hbar \to 0$, which, from the original ideas of Dirac [38] and Feynman, should be determined by the path which makes $S(\gamma)$ stationary, i.e.

according to Hamilton's principle, the trajectory of the classical motion.

Unfortunately, at this level the Feynman path integral is not a well-defined mathematical object, in particular $\mathcal{D}(\gamma)$ is not a measure. However, under the influence of the work of Feynman, M. Kac was able to prove that the solution of the heat equation can be written as an integral with respect to the Wiener measure over the space of paths [75].

During the last four decades many works have been devoted to the mathematical definition of this oscillatory integral and to its probabilistic interpretation as expectation value with respect to stochastic processes. Let us briefly say that the approaches which give a mathematical meaning to Feynman's path integral can be roughly classified in four classes. The approach via a limiting procedure, which is the most popular, is useful for explicit evaluation but it is difficult to study the mathematical properties of the limit [37,70b,76]. The oscillatory integral approach (Fresnel integral on an infinitely dimensional Hilbert space) and a method of stationary phase on Hilbert space are well-adapted to investigate asymptotical expansions around $\hbar = 0$ [3 a, 7b] The Euclidean strategy is obtained by analytical continuations to imaginary time and in this way establishes a connection between Schrödinger's equation and the heat equation, furnishing a probabilistic description in the framework of functional integration w.r.t. Wiener measure [57,100a]. As mentioned, Kac has shown that the solution of the Euclidean Schrödinger equation

$$\begin{cases} \hbar \dfrac{\partial}{\partial t} f(t,x) = \dfrac{\hbar^2}{2m} \Delta f(t,x) - V(x) f(t,x) \\[2ex] f(0,x) = \varphi(x) \end{cases} \tag{1.13}$$

can be written (Feynman-Kac formula)

$$f(t,x) = \mathbb{E}_W \left[e^{-\frac{1}{\hbar}\int_0^t V(\gamma(\tau)+x)\,d\tau} \varphi(\gamma(0)+x) \right] \tag{1.14}$$

where $\mathbb{E}_W$ denotes the expectation w.r.t. Wiener measure. Formally, this measure can be expressed as

$$dW(\gamma) = e^{-\int_0^t \frac{1}{2}\dot{\gamma}^2(\tau)\,d\tau} \mathcal{D}(\gamma). \tag{1.15}$$

This formula has been used to obtain a definition of Feynman's path integral by analytical continuation and has played an important rôle in tackling quantum field theory and the infinities of renormalization [57]. Unfortunately, this probabilistic approach cannot be directly extended to quantum mechanics in "real" time since the Wiener measure with a complex covariance is not a measure [20]. Nevertheless, as an application of generalized Brownian functionals (e.g. distributions on Wiener space of [86]) a definition of Feynman's path integral, in real time, has been proposed and discussed by L. Streit and T. Hida, see [66] and [102] and references therein. Another approach was initiated by V.P. Maslov and A.M. Chebotarev [24,28,84] and can be generalized to Feynman path integral on phase space [27]. This probabilistic framework differs from the previous one in the sense that it works directly with real time and that the involved stochastic processes are jump processes.

If a successful theoretical framework has been found, it is interesting to try to reformulate its structure in different forms in order to isolate some aspects and perhaps to find new implications and developments. The purpose of these lectures is to present physical and mathematical aspects of stochastic mechanics. As we have suggested at the beginning, the domain of physical applications of stochastic mechanics is not restricted to quantum mechanics. But before introducing the basic notions of stochastic mechanics, let us first discuss the rôle of probability theory on quantum physics.

I.2 Probabilistic Interpretation of Quantum Mechanics and Probability Theory

I.2a Historical Remarks

From the beginning of quantum mechanics the search for a stochastic interpretation was motivated by the conspicuous similarity between the free Schrödinger equation and the heat equation. One of the first to draw attention to such similarity was E. Schrödinger himself in 1931 [98b,c]. He compared the free equation for the quantum mechanical wave function in one dimension with the diffusion equation

$$\frac{\partial \rho}{\partial t} = \nu \frac{\partial^2 \rho}{\partial x^2} \tag{1.15}$$

where $\rho(t,x)$ is the density of the particle and ν the diffusion coefficient. Discussing the problem of finding the probability distribution at time t with $t \in [t_0,t_1]$ if $\rho(t_0,x)$ and $\rho(t_1,x)$ are known, he showed that $\rho(t,x)$ is given as a product of two factors $\rho_1\rho_2$, in striking analogy to the quantum mechanical expression $\psi^*\psi = |\psi|^2$ for the quantum probability density. But the differences $(\rho_1,\rho_2$ are positive and ψ is complex valued; moreover, in diffusion theory the probability density ρ itself and in quantum theory only the probability amplitude ψ are solutions of a partical differential equation) were, from Schrödinger's point of view, important enough to convince him not to try to adopt a stochastic interpretation of quantum mechanics. Using a different class of diffusion processes, the "Bernstein processes", J. C. Zambrini [114] has recently shown that the realization of the program initiated by Schrödinger gives the genuine Euclidean version of stochastic mechanics.

Not only the Schrödinger equation which can be written for free particles in one dimension

$$\frac{\partial\psi}{\partial t} = \varepsilon\frac{\partial^2\psi}{\partial x^2}\,, \qquad \varepsilon = -\frac{i\hbar}{2m} \tag{1.16}$$

has a stochastic analogue, namely the diffusion equation

$$\frac{\partial\rho}{\partial t} = \nu\frac{\partial^2\rho}{\partial x^2}\,, \qquad \nu \in \mathbb{R}_+ \,, \tag{1.17}$$

but,as demonstrated in 1933 by Fürth [51b], see also [71], an analogy also exists for the Heisenberg uncertainty relations, which are often regarded as the characteristic feature of quantum mechanics.

In this discussion of the stochastic analog of Heisenberg's relation, Fürth was the first to derive the uncertainly relation as follows. Let x_0 be the initial position of a particle and v its initial velocity, then its position at time t will be given by

$$x = x_0 + vt\,. \tag{1.18}$$

The mean square $\alpha = \langle x^2\rangle$ is therefore

$$\alpha = \langle x_0^2\rangle + 2\langle x_0 v\rangle t + \langle v^2\rangle t^2\,. \tag{1.19}$$

From $\alpha = \int_{\mathbb{R}} x^2|\psi(x,t)|^2 dx$ and using (1.16) and its complex conjugate, one obtains after partial integrations

$$\frac{d^2\alpha}{dt^2} = -8\varepsilon^2\,||\psi||_2^2$$

and

$$\frac{d^3\alpha}{dt^3} = 0$$

from which follows that α is a quadratic function of t, the coefficient $<v^2>$ of t^2 being

$$<v^2> = \frac{1}{2}\frac{d^2\alpha}{dt^2} = -4\varepsilon^2 \|\psi\|_2^2 \quad . \tag{1.20}$$

From the obvious inegality

$$\left|\frac{x}{2\alpha}\psi + \psi'\right|^2 \geq 0$$

and $\|\psi\|_2^2 = 1$ it follows immediately that

$$\|\psi'\|_2^2 \geq \frac{1}{4\alpha}$$

and hence by (1.20)

$$<x^2> <v^2> \geq -\varepsilon^2 \ .$$

Thus, setting $\Delta x = \sqrt{\alpha}$ and $\Delta p = m\sqrt{<v^2>}$, Fürth obtains the Heisenberg uncertainty relation

$$\Delta x \cdot \Delta p \geq \frac{h}{4\pi} \ . \tag{1.21}$$

Along the same lines Fürth defined the uncertainty of position for the diffusion process by

$$\beta = <x^2> = \int_{\mathbb{R}} x^2 \rho(t,x)dx$$

where ρ is a positive solution of the diffusion equation (1.17) such that $\|\rho\|_1 = 1$.

This implies $\frac{d\beta}{dt} = 2\nu$ and

$$\beta = x_0 + 2\nu t \ .$$

Hence in this case, in contrast to (1.19), the uncertainty increases linearly in time, since the motion of each particle, instead of being given as before with an initial velocity, results now from the action of random collisions with other particles. The <u>diffusion current</u> j , i.e. the amount of particles through a fixed unit area per unit time given by

$$j = -\nu \nabla \rho$$

was used by Fürth to define an <u>osmotic velocity</u> u

$$u = \frac{1}{\rho} j = - \nu \frac{\nabla\rho}{\rho} = - \nu \nabla \log \rho .$$

Using

$$\langle u^2 \rangle = \int_{\mathbb{R}} u^2 \rho(t,x) dx = \nu^2 \int_{\mathbb{R}} \frac{\rho'^2}{\rho} dx \qquad (1.22)$$

and the obvious inequality

$$\left(\frac{\rho'}{\rho} + \frac{x}{\beta}\right)^2 \geq 0 .$$

Fürth obtains, since $\|\rho\|_1 = 1$

$$\int_{\mathbb{R}} \frac{\rho'^2}{\rho} dx \geq \frac{1}{\beta}$$

and hence by (1.22)

$$\langle x^2 \rangle \langle u \rangle^2 \geq \nu^2 .$$

Now, with Δx and Δu defined as before, Fürth derives the following diffusion uncertainty relation

$$\Delta x \cdot \Delta u \geq \nu \qquad (1.23)$$

in analogy to (1.21).

However, as he pointed out, in contrast to (1.21), where the lower bound is given by the universal constant $\hbar$ originating from the disturbance produced by the measurement process itself [64], the lower bound in (1.23) associated with the random agitation of the surrounding medium can be arbitrarily small, for example by lowering the temperature since ν is proportional to T [42].

I.2b The Wigner Approach

The search for an interpretation of quantum mechanics in the framework of a classical probabilistic theory found support and encouragement between the thirties and the fourties thanks to a result obtained by E. Wigner [111], which seemed to carry more weight than mere analogy considerations. In fact, it suggested the possibility of formulating quantum mechanics in terms of phase space ensembles. Wigner introduced a function $\mathcal{W}(q,p)$ of position and momentum variables, containing all information about the quantum state. For a pure state this function integrated over momentum space, yields the quantum mechanical probability distribution of position, namely $|\psi(q)|^2$, and when integrated over the position space yields the corresponding probability distribution of momenta, namely $|\tilde{\psi}(p)|^2$, where $\tilde{\psi}$ is the Fourier transform of ψ. Moreover, if $f(q,p)$ is any classical observable, i.e. a function on phase space, the quantum mechanical expectation of the observable $Q(f)$ associated to f in the state defined by the Wigner distribution $\mathcal{W}$ is given by

$$\langle Q(f)\rangle_{\mathcal{W}} = \int_{\mathbb{R}^{2n}} f(q,p)\mathcal{W}(q,p)\, dqdp \,. \tag{1.24}$$

For pure state ψ the Wigner distribution $\mathcal{W}_\psi$ is given by the explicit formula

$$\mathcal{W}_\psi(q,p) = \frac{1}{(\pi\hbar)^n}\int_{\mathbb{R}^n} dx\ \psi^*(q+x)e^{2i\frac{p\cdot x}{\hbar}}\psi(q-x) \tag{1.25}$$

and admits the marginal distributions

$$\int_{\mathbb{R}^n}\mathcal{W}_\psi(q,p)dp = |\psi(q)|^2 \,, \quad \int_{\mathbb{R}^n}\mathcal{W}_\psi(q,p)dq = |\tilde{\psi}(p)|^2 \tag{1.26}$$

where

$$\tilde{\psi}(p) = \frac{1}{(2\pi\hbar)^{n/2}}\int_{\mathbb{R}^n} e^{i\frac{p\cdot x}{\hbar}}\psi(x)dx \,.$$

Hence the Wigner distribution appears as a kind of density on the phase space. However, although $\mathcal{W}$ is real, it is not positive and consequently has not quite the interpretation of a statistical distribution in classical physics. It is natural to ask whether the fact that the joint probability distribution for coordinates and momenta can be relaxed. More precisely, two types of questions can be investigated.

(i) Is it possible to describe a quantum mechanical state in terms of an average over stochastic processes?

(ii) Is it possible to reformulate quantum mechanics in a purely probabilistic framework so that observables are represented by random variables?

During the last five decades these questions have attracted the attention of numerous mathematicians as well as physicists. The first question is closely connected with the mathematical definition of Feynman's path integral. Indeed, the Wigner description provides a good candidate to give a functional integral representation of the quantum mechanical state in the sense that the time evolution of the Wigner function can be described using stochastic processes with values in the product of the phase space and the torus [27]. This approach gives a probabilistic description on an extended space. From a mathematical point of view the property that the Wigner distribution is given as the difference of two positive distributions implies the existence of stochastic processes valued in the product of the phase space and the torus (or disc)[85c]. The second question, relative to the possibility of representing quantum observables by random processes, was studied first by E. Moyal [88].

I.2c Probabilistic Description of Commuting Observables

To be more precise, let us recall some basic facts about quantum mechanics. For more details see e.g. [91]. The main mathematical structure introduced in quantum mechanics is the superposition of states and algebraic operations on observables. To each quantum mechanical system there exists a Hilbert space $\mathcal{H}$ such as there is a unit ray $\underline{\psi} = \{e^{i\theta}\psi,\ 0 \leq \theta < 2\pi\}$, $||\psi||_1 = 1$, corresponding to each state ψ and a self-adjoint operator A_{op} corresponding to each observable A. We need an axiom stating that there are sufficiently many $\underline{\psi}$ and A_{op}. As an extreme case we have: $\underline{\psi}$ exhausts all unit rays on $\mathcal{H}$ and the set of A_{op} contains all projection operators. The evolution in time of the system is described by a one parameter family of unitary operators U_t on $\mathcal{H}$ and can be achieved in two ways. In the Schrödinger picture the state of the system evolves in time according to

$$\psi(t) = U_t\,\psi_o$$

where ψ_o is the initial state at time 0 and the observables do not change with time. It must be remarked that the Wigner description of the state gives rise to a Schrödinger picture and the time evolution of the Wigner function obeys a Schrödinger type equation. In the Heisenberg picture the observables evolve in time according to

$$A_{op}(t) = U_t^{-1} A_{op}\, U_t$$

and the state does not change with time.

If the system is in the state ψ and if A is an observable, there exist projection-valued measures $\{E_\lambda\}$ on $\mathcal{H}$ such as

$$A_{op} = \int_{\mathbb{R}} \lambda \, dE_\lambda \,.$$

Then $\langle\psi,E_\lambda\psi\rangle$ is the probability that, if we perform an experiment to determine the value of the observable A in the state ψ, we obtain a result smaller than or equal to λ since

$$\langle A_{op}\rangle_\psi = \langle\psi,A_{op}\psi\rangle = \int_{\mathbb{R}} \lambda\, d\, \langle\psi,E_\lambda\psi\rangle \tag{1.27}$$

is the expectation value of A in the state ψ .

The observable A has the value λ with probability one, if ψ is an eigenvector of A_{op} associated with the eigenvalue λ

$$A_{op}\psi = \lambda\psi \ .$$

This is the way in which probabilistic ideas play a central rôle in quantum mechanical theory. Hence given the state ψ the observable A may be regarded as a random variable $(\Omega,\mathcal{B},P^A(d\lambda))$ with $\Omega = \mathbb{R}$, $\mathcal{B}$ the usual Borel field, and

$$P_\psi^A = \langle\psi,dE_\lambda\psi\rangle \quad .$$

Similarly, any number of commuting self-adjoint operators can be regarded as random variables on a appropriate probability space. But not all observables of a quantum system are commuting and it follows from a theorem of von Neumann that the set of all observables in a given state cannot be regarded as a family of random variables on a probability space. More precisely, we have the following

Theorem (1.3) (Nelson-von Neumann [90b,91])

Let $A_1,\ldots,A_n$ be n self-adjoint operators on a Hilbert space H , such that for all $x \in \mathbb{R}^n$ the operator $x.A$ defined by

$$x.A = \sum_{i=1}^{n} x_i A_i$$

is essentially self-adjoint. Then, either the A_i commute or there exists $\psi \in H$ with $||\psi|| = 1$ such that it is not possible to find random variables $\xi_1,\ldots,\xi_n$ on a probability space $(\Omega,\mathcal{B},P)$ with the property that for all $x \in \mathbb{R}^n$ and all $\lambda \in \mathbb{R}$

$$P_\psi(x\cdot\xi\geq\lambda) = \langle\psi,E_\lambda(x.A)\psi\rangle$$

where $x\cdot\xi = \sum_{i=1}^{n} x_i\ \xi_i$ and $\{E_\lambda(x.A)\}$ are the projection-valued measures associated with the closure of $x.A$.

Remark

Translating this result into the setting of quantum mechanics one is led to say that n observables may be regarded as random variables in all states if and only if they commute.

Proof

We don't distinguish notationwise between $x.A$ and its closure. Suppose that for each unit vector ψ in H there is such an n-uple ξ of random variables, and let μ_ψ be the probability distribution of $\xi \in \mathbb{R}^n$. That is, for each Borel set B in $\mathbb{R}^n$, $\mu_\psi(B) = P_\psi(\xi \in B)$.

Let us compute the characteristic function of the measure μ_ψ

$$\int_{\mathbb{R}^n} e^{ix\cdot\xi} d\mu_\psi(\xi) = \int_{-\infty}^{+\infty} e^{i\lambda} dP_\psi(x\cdot\xi \geq \lambda)$$

$$= \int_{-\infty}^{+\infty} e^{i\lambda} <\psi, dE_\lambda(x.A)\psi>$$

$$= <\psi, e^{ix.A}\psi>.$$

Thus, the measure μ_ψ is the Fourier transform of $<\psi, e^{ix.A}\psi>$. By the polarization identity, if φ and ψ are in H there is a complex measure $\mu_{\varphi\psi}$ which is the Fourier transform of $<\varphi, e^{ix.A}\psi>$ and verifies $\mu_{\psi\psi} = \mu_\psi$. For any Borel set $B \subset \mathbb{R}^n$ by the Riesz theorem there exists a unique operator $\mu(B)$ such that $<\varphi, \mu(B)\psi> = \mu_{\varphi\psi}(B)$.

Thus we have

$$\int_{\mathbb{R}^n} e^{ix\xi} <\varphi, d\mu(\xi)\psi> = <\varphi, e^{ix.A}\psi>$$

the operator $\mu(B)$ is positive since μ_ψ is a positive measure. Consequently, if we have a finite set of elements $\psi_j \in H$ and corresponding points $x_j \in \mathbb{R}^n$, then

$$\sum_{j,k} <\psi_k, e^{i(x_j-x_k)\cdot A}\psi_j>$$

$$= \sum_{j,k} \int_{\mathbb{R}^n} e^{i(x_j-x_k)\cdot\xi} <\psi_k, d\mu(\xi)\psi_j>$$

$$= \int_{\mathbb{R}^n} <\psi(\xi), d\mu(\xi)\psi(\xi)> \geq 0$$

if we define

$$\psi(\xi) = \sum_j e^{ix_j \cdot \xi} \psi_j(\xi) \quad .$$

Furthermore, $e^{i0.A} = 1$ and $e^{i(-x).A} = (e^{ix.A})^*$.

Under these conditions, the theorem on unitary dilation of Nagy [93] implies that there is a Hilbert space K containing H and a unitary representation $x \to U(x)$ of $\mathbb{R}^n$ on K such that, if $\mathbb{P}$ is the orthogonal projection of K onto H, then

$$\mathbb{P}\, U(x)\psi = e^{ix.A}\psi$$

for all $\mathbb{R}^n$ and all $\psi \in H$.

Since $e^{ix.A}$ is already unitary,

$$||U(x)\psi|| = ||e^{ix.A}|| = ||\psi||$$

so that $$||\mathbb{P}U(x)\ \psi|| = ||U(x)\psi|| \ .$$

Consequently, $\mathbb{P}\,U(x)\psi = U(x)\psi$ and each $U(x)$ maps H into itself, so that $U(x) = e^{ix.A}$ for all $\psi \in H$. Since $x \to U(x)$ is a unitary representation of the commutative group $\mathbb{R}^n$, the $e^{ix.A}$ all commute, and consequently the A_j commute.

Remark 1

Nelson-von Neumann's theorem does not claim that there are no state having a joint probability distribution for non-commuting self-adjoint operators but asserts only that such states are exceptional.

Remark 2

This result means that if quantum mechanics has a probabilistic interpretation, it is not a probability theory in the classical mathematical sense. To solve in a rather natural way some old problems concerning the interpretation of quantum mechanics and its mathematical foundations, a non-commutative stochastic calculus has been developed recently see [67], [89d,e]. In this framework one considers a non-commutative stochastic process (or quantum process) with values in Fock space, which is a candidate to obtain a Feynman path integral description of Bose and Fermi quantum field theory, (see also P. Garbaczenski [54]).

I.3 Jacobi and Madelung Fluid

While Schrödinger's attempt to give an interpretation of quantum mechanics found its support primarily in the analogy of wave phenomena, the similarity of the equation for the wave function with the equations describing hydrodynamical flows formed the basis for another early attempt to account for quantum mechanical phenomena in the framework of the classical physic of continuous media. In 1926 E. Madelung [83] gave an hydrodynamical description of quantum mechanics very reminiscent of the Hamilton-Jacobi fluid. Since this approach will be useful to understand the general structures of stochastic mechanics, let us give a short review of the Hamilton-Jacobi description of classical mechanics and of the Madelung fluid. For more details see [60a].

I.3a The Hamilton-Jacobi Fluid

We consider a classical dynamical system with n degrees of freedom and with phase space $\mathbb{R}^n \times \mathbb{R}^n$. We denote by $(x,y) \in \mathbb{R}^n \times \mathbb{R}^n$ a point in phase space and by $(q(t), p(t))$ a path in phase space as function of time. If $H(x,y,t)$ is the (smooth) Hamilton function, then the canonical Hamilton equation of motion can be written as

$$\begin{aligned} \dot{q}(t) &= (\nabla_y H)(q(t),p(t),t) \\ \dot{p}(t) &= -(\nabla_x H)(q(t),p(t),t). \end{aligned} \tag{1.28}$$

Given the initial values at time t_o,

$$q(t_o) = x_o \;,\; p(t_o) = y_o \;, \tag{1.29}$$

the integration of (1.28) gives the path in phase space

$$q(t) = q(x_o,y_o,t), \quad p(t) = p(x_o,y_o,t) \;. \tag{1.30}$$

In classical mechanics time reversal plays an important rôle. Let

$$T: \mathbb{R} \times \mathbb{R}^n \times \mathbb{R}^n \to \mathbb{R} \times \mathbb{R}^n \times \mathbb{R}^n$$

be defined by $T(t,x,y) = (t',x',y')$ with

$$t' = -t,\; x' = x,\; y' = -y$$

and assume that the Hamilton function is invariant under time reversal

$$H(x',y',t') = H(x,y,t)$$

Using (1.28) and their transform under T it follows that, if (1.30) is a solution of Hamilton's equation, then

$$q'(t') = q(t), \; p'(t') = -p(t)$$

is a solution of the equation of motion (t' playing now the rôle of time) but with the following initial condition

$$q'(t_o') = x_o, \; p'(t_o') = -y_o \; .$$

Now, let us introduce the Hamilton-Jacobi equation

$$\frac{\partial S}{\partial t} + H(x,\nabla S,t) = 0 \tag{1.31}$$

supplemented by the initial condition

$$S(x,t) = S_o(x) \quad x \in \mathbb{R}^n \; . \tag{1.32}$$

If some S satisfying (1.31) and (1.32) can be found, then it is well-known that it is possible to obtain a solution of Hamilton's equations, by only specifying the initial position $q(t_o) = x_o$. From this point of view each given S defines a whole family of paths on phase space, each of which characterized by the initial position alone, as we shall demonstrate now.

Starting with a given S let us introduce the momentum field p by

$$p(x,t) = \nabla S(x,t) \tag{1.33}$$

and the velocity field v by

$$v(x,t) = (\nabla_y H)(x, p(x,t),t) \tag{1.34}$$

consider now the first order differential system

$$\begin{cases} \dot{q}(t) = v(q(t),t) \\ q(t_o) = x_o \end{cases} \tag{1.35}$$

and suppose that $q(t)$ is solution of (1.35). Setting

$$p(t) = p(q(t),t) = \nabla S(q(t),t) \tag{1.36}$$

we see easily that $(q,(t),p(t))$ defines a path in phase space, along which the Hamilton equations are satisfied.

Instead of considering each single trajectory specified by $q(t_o) = x_o$ for some given S, we can also consider a continuous distribution of paths, associated with some density $\rho(x,t)$ in configuration space.

In fact, starting from the Liouville equation for the density $\tilde{\rho}(x,y,t)$ in phase space

$$\frac{\partial\tilde{\rho}}{\partial t} + \{\tilde{\rho},H\} = 0 \tag{1.37}$$

where $\{\cdot,\cdot\}$ denotes the Poisson brackets, i.e.

$$\{a,b\} = \nabla_x a.\nabla_y b - \nabla_y a.\nabla_x b \ . \tag{1.38}$$

We can make the following Ansatz

$$\tilde{\rho}(x,y,t) = \rho(x,t)\delta(y-\nabla S(x,t)) \tag{1.39}$$

wich constrains the momentum to verify (1.33) for all times. It is now easy to see that $\tilde{\rho}$ satisfies (1.37), if the density ρ is solution of the continuity equation

$$\frac{\partial\rho}{\partial t} + \nabla(\rho v) = 0 \tag{1.40}$$

where v is defined by (1.34).

We define the Hamilton-Jacobi fluid as a mechanical system living a configuration space, described by two fields, the Hamilton-Jacobi function $S(x,t)$ and the density $\rho(x,t)$ and evolving in time in such a way that the Hamiltonian-Jacobi equation

$$\frac{\partial S}{\partial t} + H(x,\nabla S,t) = 0 \tag{1.41}$$

is satisfied.

It must be remarked that, vice-versa, if we have a fluid verifying (1.41), then the described procedure will provide a particle interpretation.

In the simplest case, the Hamilton function is of the form

$$H(x,y) = \frac{y^2}{2m} + V(x) \tag{1.42}$$

and the Hamilton equations are

$$\dot{q}(t) = \frac{p(t)}{m} \quad , \quad \dot{p}(t) = - \nabla V(q(t)) \ . \tag{1.43}$$

Therefore, the Hamilton-Jacobi fluid is described by the system

$$\begin{cases} \frac{\partial S}{\partial t} + \frac{1}{2m} (\nabla S)^2 + V(x) = 0 \\ \frac{\partial \rho}{\partial t} + \nabla(\rho v) = 0 \ . \\ v(x,t) = \frac{1}{m} \nabla S(x,t) \ . \end{cases} \tag{1.44}$$

Let us emphasize that the possibility of giving a particle picture for hydrodynamical equations, as the previous one based on deterministic paths, comes from the fact that the equation for S is decoupled from the equation for the density ρ. In fact, the equation for ρ is simply a continuity equation, expressing the local conservation of mass, while the equation for S, which is of first order in time, allows the introduction of particle trajectories identified with its characteristic lines, as well-known from the general theory of first order partial differential equation, see e.g. [9].

I.3b The Madelung Fluid

As remarked by Madelung [83] just at the beginning of wave mechanics, it is also possible to reformulate the standard Schrödinger equation associated with the Hamilton function (1,42) $H = \frac{y^2}{2m} + V(x)$ which is namely

$$i\hbar \frac{\partial \psi}{\partial t} = -\frac{\hbar^2}{2m} \Delta\psi + V\psi \tag{1.45}$$

with $\psi \in L^2(\mathbb{R}^n, dx)$ and Δ the Laplace-operator in $\mathbb{R}^n$.

Let us separate modulus and phase in the wave function

$$\psi(x,t) = e^{R(x,t) + \frac{i}{\hbar} S(x,t)} \tag{1.46}$$

where R and S are real-valued functions.

Substituting expression (1.46) for the Schrödinger equation and separating real and imaginary parts, we obtain for the imaginary part

$$\frac{\partial R}{\partial t} = - \frac{1}{2m} (\Delta S + 2\nabla R \cdot \nabla S) \ . \tag{1.47}$$

Introducing the vector field

$$v(x,t) = \frac{1}{m} \nabla S(x,t) \tag{1.48}$$

and the probability density

$$\rho(x,t) \equiv |\psi(x,t)|^2 = e^{2R(x,t)} \tag{1.49}$$

expression (1.47) can be put in the form of a continuity equation

$$\frac{\partial \rho}{\partial t} + \nabla(\rho v) = 0 . \tag{1.50}$$

On the other hand, we obtain for the real part

$$\frac{\partial S}{\partial t} = \frac{\hbar^2}{2m}[\Delta R + (\nabla R)^2 - (\frac{\nabla S}{\hbar})^2] - V. \tag{1.51}$$

The two equations (1.50) and (1.51) are strictly equivalent to the Schrödinger equation. Using now

$$\nabla(e^R) = (\nabla R)e^R$$
$$\Delta e^R = [\Delta R + (\nabla R)^2]e^R ,$$

(1,51) can be put in the form

$$\frac{\partial S}{\partial t} + \frac{1}{2m}(\nabla S)^2 + V - \frac{\hbar^2}{2m}\frac{\Delta e^R}{e^R} = 0 . \tag{1.52}$$

We call the hydrodynamical system described by (1.48), (1.50) and (1.52) the Madelung fluid. The analogy with the Hamilton-Jacobi fluid is striking, but now no immediate particle interpretation is available.

This is due to the mysterious nature of the "quantum mechanical potential"

$$V_{QM} = - \frac{\hbar^2}{2m}\frac{\Delta e^R}{e^R} \tag{1.53}$$

which depends on the density $(e^R = \sqrt{\rho})$.

In the next section, following E. Nelson, we will show that a natural and straightforward particle interpretation of the Madelung fluid is indeed possible, but only by allowing a random character to the underlying paths.

The classical approximation consists in setting $\hbar$ equal to zero in (1.52) in which the randomness disappears and the trajectories become those of the classical theory, the Madelung fluid reducing to the Hamilton-Jacobi fluid.

We emphasize the fact that we have spoken about a particle picture and not claimed that particles and trajectories really exist in the physical realm.

<u>Remark</u>

In the classical approximation, an optical analogy is even more suggestive than this hydrodynamical analogy, especially for stationary solutions. Since the velocities of the particles are proportional to ∇S, the trajectories of these particles are orthogonal to the surfaces of equal phase $S = \text{const}$. In the language of optics, the latter are the wave fronts and the trajectories of the particles are the rays. Hence, the classical approximation is equivalent to the geometrical optics approximation.

II. KINEMATICS OF STOCHASTIC DIFFUSION PROCESSES

II.1 Brownian Motion

Let us consider a system such that its state at each time is completely specified by the position of a point ξ in the configuration space $\mathbb{R}^d$ and assume a deterministic evolution equation

$$\dot{\xi}(t) = b(\xi(t),t) \tag{2.1}$$

where $b(x,t)$ is some assigned vector field in $\mathbb{R}^d$ which may depend on external force fields acting on the system. Under appropriate regularity conditions on b and for given initial condition

$$\xi(t_o) = \xi_o \tag{2.2}$$

(2.1) has, at least locally, a unique solution. An example of such deterministic evolution, of the previous form, is given by the Hamilton equations (1.28).

In many case there is no complete control on the external forces or the equation of motion (2.1) is derived from an approximation in which some degrees of freedom have been neglected, (e.g. the motion of a small particle in a fluid where we neglect the interaction with the molecules of the fluid). A wide class of physical phenomena, ranging from statistical physics to astrophysics and control theory, shows that very often a suitable modification of the dynamics gives good results, provided we take into account all effects coming from external fields and neglected degrees of freedom, by introducing some phenomenologically determined random disturbance acting on the system during the evolution.

To make the previous description more precise, let us consider the motion of a particle in $\mathbb{R}^3$ submitted to rapidly varying forces due to the collisions with molecules of the surrounding fluid, and let us make the assumption that the external gravitational field is negligible. The particle moves because it is constantly buffeted by the molecules. Taking some reasonable assumptions about the behaviour of the collisions, one can deduce the probability of finding the particles in any given subset of $\mathbb{R}^3$ at time t, knowing that it was located at $x \in \mathbb{R}^3$ at initial time $t = 0$. That is just what was done by A. Einstein in his description of the Brownian motion as we have seen in section (I.1.a).

We will first give an idea on how the Brownian motion process can be obtained from a random walks. The particle in motion moves of course in a three dimensional space but we can think of its projection on a coordinate axis. We describe now a one dimensional random walk which in a certain limiting case possesses the properties of Brownian motion. Since numerous impacts are received per second, we will shorten the unit of time, but we must shorten the unit of length so as to gain the correct model. Let τ be the new time-unit, in other words, τ is the time between two successive collisions. Thus t/τ steps are made by the particle in (old) time t. Each step is a symmetrical Bernoulli random variable and we suppose now that the step is of magnitude proportional to $\sqrt{\tau}$, namely $\sigma\sqrt{\tau}$, for each step k

$$P[\xi_k = \sigma\sqrt{\tau}] = P[\xi_k = -\sigma\sqrt{\tau}] = \frac{1}{2}$$

we have then

$$\mathbb{E}[\xi_k] = 0$$

$$\mathrm{Var}(\xi_k) = \mathbb{E}[\xi_k^2] = \frac{1}{2}(\sigma\sqrt{\tau})^2 + \frac{1}{2}(-\sigma\sqrt{\tau})^2 = \sigma^2\tau$$

where $\mathbb{E}$ denotes the expectation value.
Let $X_0 = 0$ then the position at time t (or after t/τ steps) is just

$$X_t = \sum_{k=1}^{t/\tau} \xi_k \;.$$

If τ is much smaller than t, t/τ is large and may be thought of as an integer. Hence we have

$$\mathbb{E}[X_t] = 0$$

$$\mathrm{Var}\,[X_t] = \frac{t}{\tau}\sigma^2\tau = \sigma^2 t \;.$$

Furthermore, if t is fixed and $\tau \to 0$, then by the De Moivre-Laplace central limit theorem X_t will have the normal distribution $N(0,t)$. Hence if in our model, in which the random particle jumps over a distance $\pm\,\sigma\sqrt{\tau}$ with equal probability in time τ, we perform the limit $\tau \to 0$, the limiting process is the Brownian motion W_t (also called Wiener process).

In this model we have used the fact that the displacement is the sum of many very small independent contributions; this assumption is a good approximation. Indeed, if relatively high viscosity is assumed, so that the velocity of the particle is very quickly damped, the displacements in non-overlapping intervals of time should be independent. By symmetry (homogeneity and isotropy) $\mathbb{E}[\xi_t] = 0$ and if the physical conditions (temperature, pressure, etc.) remain constant, we should have that $\mathbb{E}[(\xi_{t+s} - \xi_t)^2] = f(s)$ is independent of t. This condition with the independence implies $f(s_1 + s_2) = f(s_1) + (s_2)$. The assumption that f is continuous leads to $f(s) = cs$ and the variance of increments is linear in time.

If we denote by dt a strictly positive interval of time (small compared to the time needed by the particle to cover a macroscopic distance and large with respect to the interaction time τ), and denoting by W_t the corresponding limiting process, in three dimensions we have

$$E[dW_t] = 0 , \quad \mathbb{E}[dW_t^i \, dW_t^j] = \sigma^2 \delta^{ij} dt . \tag{2.3}$$

The transition probability of a Gaussian random variable is completely specified by its mean and covariance. Therefore, (2.3) gives us the transition function $p(x,t,y,s)$ for the d-dimensional Brownian motion

$$p(x,t,y,s) = \frac{1}{(2\pi\sigma^2(s-t))^{d/2}} \, e^{-\frac{|y-x|^2}{2\sigma^2(s-t)}} \tag{2.4}$$

for any finite time $t < s$.

The physical meaning of p is that $p(x,t,y,s)dy$ represents the probability that the particle (performing Brownian motion) will reach at time s the Borel set dy around y if it started from x at time $t \leq s$.

Clearly, we have the initial condition

$$\lim_{s \downarrow t} p(x,t,y,s) = \delta\,(x-y) \tag{2.5}$$

and p satisfies the parabolic equation <u>(Heat equation)</u>

$$\frac{\partial p}{\partial s} = \frac{\sigma^2}{2} \Delta_y p . \tag{2.6}$$

The coherence of the physical interpretation of p is due to the normalization condition

$$\int_{\mathbb{R}^3} p(x,t,y,s)\,dy = 1 \tag{2.7}$$

and

$$p(x,t,y,s) = p(0,0,y-x,s-t) = \underline{p}(y-x,s-t) . \tag{2.8}$$

This property is the consequence of the space-time translation invariance of our model.

<u>The Chapman-Kolmogorov equations</u> (compatibility conditions)

$$p(x,t,y,s) = \int_{\mathbb{R}^d} p(x,t,x',t')p(x',t',y,s)\,dx' \tag{2.9}$$

for any t' such that $t \le t' \le s$, takes care of the fact that at each time the process develops without remembering the positions occupied at an earlier time; this is the <u>Markov property</u>. The physical significance of (2.9) is obvious. The probability that the particle which is located in x at time t was in y at time s is the sum over all the possible intermediate state x' at time t', $t \le t' \le s$ of the product of the probability that the particle moves from x at time t to y' at some intermediate time t' and of the probability to move from x' at time t' to y at time s.

Now, if we denote by W_t the random position of the particles at time t, then the knowledge of the transition probability (2.4) allows us to give meaning to all <u>correlation functions</u> of the process provided the initial distribution $\rho(x,t_o)$, at some initial time t_o for W_{t_o} is known

$$\mathbb{E}\,[F(w_o, \ldots, w_{t_n})] = \int F\,(x_o, \ldots, x_n)p(x_o,t_o,x_1,t_1) \\ \ldots\ldots\ldots\ p(x_{n-1},t_{n-1},x_n,t_n)\rho(x_o,t_o)\,dx_o \ldots\ldots dx_n \tag{2.10}$$

for $t_o \le t_1 \le t_2 \ldots \le t_n$.

In particular, for the density at time t we have

$$\rho(x,t) = \int p(x_o,t_o,x,t)\rho(x_o,t_o)\,dx_o \tag{2.11}$$

so that $\rho(x,t)$ satisfies also the diffusion equation

$$\frac{\partial\rho}{\partial t} = \frac{\sigma^2}{2}\,\Delta_x \rho \tag{2.12}$$

but with initial condition given by

$$\lim_{t \downarrow t_o} \rho(x,t) = \rho(x,t_o) \; . \tag{2.13}$$

Although the transition probability p and the density ρ satisfy the same equation (2.6) and (2.12) there is a deep difference between the physical meaning of p and ρ . The link between p and ρ is given through (2.11) and the initial condition (2.5) and (2.13). In particular, the knowledge of the transition probability p does not determine the density ρ which in fact depends on some specified initial condition.

If we are interested in the behaviour of the random trajectories $t \to W_t$ of the Brownian particle, we need to know more than the transition function (and correlation functions). We have to prove the existence of a probability measure on the space of trajectories consistent with the Brownian transition function in the following sense: for any finite collection of measurable sets $A_1,\dots,A_n$ in $\mathbb{R}^d$ and any $0 < t_1 < t_2 \dots < t_n$ we have

$$\begin{aligned} &P(W_{t_1} \in A_1 \,, \dots , W_{t_n} \in A_n) \\ &= \int_{A_1} dx_1 \dots \int_{A_n} dx_n \; p(x,o,x_1,t_1)p(x_1,t_1,x_2,t_2) \dots \\ &\qquad \dots p(x_{n-1},t_{n-1},x_n,t_n) \end{aligned} \tag{2.14}$$

where we have assumed that the path starts at the point x at time zero.

This construction was first achieved by N. Wiener in 1923 [111] and the probability measure, $P_{W,x}$, is the so-called Wiener measure (started at point x) .

Let us indicate briefly the main steps of the construction of the probability space.

Let $T = [0,t]$ be a finite interval of time and consider $\dot{\mathbb{R}}^d$ the one point compactification of $\mathbb{R}^d$. The space of trajectories Ω is the space of all functions from T into $\dot{\mathbb{R}}^d$,

$$\Omega = \{\omega | T \to \dot{\mathbb{R}}^d\} = [\dot{\mathbb{R}}^d]^T .$$

By the Tychonov theorem, it is a compact space in the product topology.

Let $C(\Omega)$ be the space of continuous functions on Ω and denote by $C_c(\Omega)$ the subset of cylindrical functions. A function $F:\Omega \to \mathbb{C}$ belongs to $C_c(\Omega)$ if there exists $k > 0$, a continuous function f from $[\dot{\mathbb{R}}^d]^k \to \mathbb{C}$ and $0 \leq t_1 < t_2 \ldots < t_k \leq t$ such that

$$F(\omega) = f\,(\omega(t_1), \ldots., \omega(t_k)), \forall\omega \in \Omega . \tag{2.15}$$

The set of cylindrical functions is a sub-*-algebra of the algebra $C(\Omega)$, and seperate the points of Ω, thus $C_c(\Omega)$ is dense in $C(\Omega)$ for the uniform topology.

Using the transition probability (2.4) and (2.9), we can define a positive form I_x on $C_c(\Omega)$ by

$$I_x(F) = \int dx_1 \ldots \int dx_k\; p(x_1,t_1,x_2,t_2)\ldots p(x_k,t_k,x,t)f(x_1,\ldots,x_k) \tag{2.16}$$

where f is a function associated with F along (2.15). It is a bounded linear functional on $C_c(\Omega)$

$$|I_x(F)| \leq \|F\|_\infty = \sup_{\omega\in\Omega} |F(\omega)|,\; F \in C_c(\Omega)$$

then I_x can be extended to a bounded linear form $\bar{I}_x$ (with the same norm as I_x) on $C(\Omega)$, by the theorem of Stone-Weierstrass. Finally, by the Riesz representation theorem $\bar{I}_x$ defines a probability measure $P_{W,x}(d\omega)$ on Ω

$$\bar{I}_x(F) = \int_\Omega P_{W,x}(d\omega)F(\omega) . \tag{2.17}$$

Hence, the rough statement (2.3) takes a precise mathematical content.

Let $\mathbb{E}$ and $\mathbb{E}[\cdot|\cdot]$ denote respectively the expectation and conditional expectation; we have

$$\begin{aligned} \mathbb{E}[dW_t | W_t] &= 0 \\ \mathbb{E}[dW_t^i\, dW_t^j | W_t] &= \sigma^2 \delta^{ij} dt \end{aligned} \tag{2.18}$$

The expectation being taken with respect to the Wiener measure $P_{W,x}$. Higher order conditional moments are of order $o(dt)$.

We have established the existence of the probability measure but it lives in a rather big space. In application, one needs a "nice" collection of $\mathbb{R}^d$-valued functions, a "large" σ-algebra on this family and a probability density concentrated on this collection. Hence it is natural to ask: where does the measure $P_{W,x}$ have its support? We have constructed $P_{W,x}$ so that it is a measure on the set of all continuous real-valued functions ω on $[0,t]$ such that $\omega(0) = x$, in the following we assume $x = 0$ (we denotes $P_W = P_{W,0}$). The limitation to continuous functions was suggested since we believe intuitively that Brownian paths are continuous. If one reason to investigate the support is to find the analytical properties of Brownian trajectories, another one is of more technical nature but very important. Suppose that we set

$$M_t(\omega) = \sup_{\tau\in[0,t]} |W_\tau(\omega)| .$$

We are interested in $M_t(\omega)$ because $P_W(\omega | M_t(\omega) \leq \delta)$ gives us the probability that the path will not be at a distance larger than δ before time t , which is a way to describe the "kinematics" of Brownian motion. The trouble is that there is no reason for $M_t(\cdot)$ to be measurable since it is the supremum over uncountably many random variables and in elementary measure theory we are only guaranteed that the supremum of countably many measurable functions is measurable. On the other hand, let $\widetilde{M}_t(\omega) = \sup_{t_n} |W_{t_n}(\omega)|$ where $\{t_n\}$ is a countable dense set in $[0,t]$, then $M_t(\omega)^n$ is measurable and $\widetilde{M}_t$ and M_t agree on the continuous functions. Therefore, the previous choice of support for the probability measure P_W as the set of continuous functions such that $f(0) = 0$ assures the measurability of M_t .

Now, what sort of condition should guarantee that paths are continuous? The condition should say roughly that if $|t-s|$ is small then the distribution of $W(t) - W(s)$ should be concentrated close to the origin. An example of such a sufficient condition is the requirement that for some $\alpha,\beta > 0$

$$\mathbb{E}(|W_t - W_s|^\beta) \leq M|t-s|^\alpha \tag{2.19}$$

which is a Hölder condition.

Let us consider, in the one dimensional case, the joint distribution of W_t and W_s which is explicitly given by

$$\frac{1}{\sqrt{2\pi\sigma^2 s}\,\sqrt{2\pi\sigma^2|t-s|}}\, e^{-\frac{x^2}{2\sigma^2 s}}\, e^{-\frac{(y-x)^2}{2\sigma^2|t-s|}}\, dx\, dy .$$

Introducing the new variables

$$\xi = \frac{x+y}{2}, \quad \eta = y-x$$

we find that the distribution of $W_t - W_s$ is

$$\frac{1}{\sqrt{2\pi\sigma^2|t-s|}}\, e^{-\frac{\eta^2}{2\sigma^2|t-s|}}\, d\eta$$

from which follows that

$$\mathbb{E}(|W_t - W_s|^4) = \frac{1}{\sqrt{2\pi\sigma^2|t-s|}} \int_{-\infty}^{+\infty} \eta^4\, e^{-\frac{\eta^2}{2\sigma^2|t-s|}}\, d\eta = 3\sigma^4|t-s|^2 . \tag{2.20}$$

More generally, because $W_t - W_s$ is a Gaussian variable, we have

$$\mathbb{E}(|W_t - W_s|^{2n}) = C_n|t-s|^n \tag{2.21}$$

where

$$C_n = (2n-1)!!\ \sigma^{2n}$$

and $\qquad (2n-1)!! = 1.3.5.....(2n-1) .$

Hence the condition is satisfied and almost all paths are continuous. Moreover, by the Kolmogorov lemma W_t has a Hölder continuous version; see e.g. [100,a].

<u>Lemma (2.1):</u> [Kolmogorov]

Let $\{\xi_t\}_{0\leq t\leq T}$ be a stochastic process obeying

$$\mathbb{E}(|\xi_{t+h} - \xi_t|^p) \leq K|h|^{1+r}$$

for some K, some $r < p$ and all t with $0 \leq t < t+h \leq T$. For $0 < q < r/p$, we have then for all dyadic rational t and s

$$|\xi_t - \xi_s| \leq C(\xi)|t-s|^q$$

where $C(\xi)$ is finite almost everywhere.

In particular, ξ has a Hölder continuous version (i.e. one can find ξ_t defined for all points with the correct joint distributions and for which the previous inequality holds a.e. for all t and s).

Then from (2.21) for $n > 1$, choosing q close to $\frac{1}{2} - \frac{1}{n}$ so by taking n large, we can obtain q's arbitrarily close to $\frac{1}{2}$. Hence the trajectories of the Wiener process are Hölder continuous of order $\alpha < \frac{1}{2}$. In fact, much more is known about the Brownian paths, see e.g. [68] . In particular, we have the following theorem.

Theorem (2.2): Almost all sample paths $\omega(t) = W_t(\omega)$ of a Wiener process are nowhere differentiable.

Let us first prove that at time $t \to \omega(t)$ is non-differentiable with probability one.

Lemma (2.3):

Let t be a fixed real number in the interval of time $[0,1]$ and $\mathcal{D}_t = \{\omega | \omega(t)$ is differentiable at time $t\}$, where $\omega(t) = W_t(\omega)$. Then the probability $P(\mathcal{D}_t) = 0$.

The measurability of $\mathcal{D}_t$ follows from the following observation. A continuous function f is differentiable at t if and only if $\lim_{h\to 0} \frac{f(t+h)-f(t)}{h}$, for rational number h, exists. If f is differentiable at point t there exists $\varepsilon > 0$ and an integer $M \geq 1$ such that for rational h such that $0 < |h| < \varepsilon$

$$|f(t+h) - f(t)| < M|h| .$$

Then

$$D_t = \bigcup_{M \geq 1} \mathcal{D}_t^M$$

where

$$\mathcal{D}_t^M = \{\omega \,|\, \exists\varepsilon > 0 \text{ s.t. } |\omega(t+h) - \omega(t)| < M|h|, 0 < |h| < \varepsilon, \forall h \in \mathbb{Q}\}$$

and

$$P(\mathcal{D}_t^M) \leq 2 \inf_h \int_0^{M\sqrt{|h|}} \frac{1}{(2\pi\sigma^2)^{1/2}} e^{\frac{y^2}{2\sigma^2}} dy = 0 .$$

Hence

$$P(\mathcal{D}_t) = 0 .$$

The lemma shows that $P(\mathcal{D}_t) = 0$.

The proof of theorem 2.2 needs a stronger result namely that the probability $P(\bigcup_{t\in[0,1]} \mathcal{D}_t)$ that $\omega(t)$ has a derivative somewhere in $[0,1]$ vanishes. Let $\omega(t)$ be differentiable at some point $t \in [0,1]$. For sufficiently large integer n , set $i = [n\,t] + 1$, where $[n\,t]$ is the integer part of $n\,t$, and let j run over $i+1$, $i+2$, $i+3$ successively, then

$$|\omega(\tfrac{j}{n}) - \omega(\tfrac{j+1}{n})| < \frac{7M}{n} \text{ , } j = i+1 \text{ , } i+2 \text{ , } i+3 \text{ .}$$

Let $D_{j,i}^{M,n}$ be the set of ω satisfying this property. Obviously, $D_{i,j}^{M,n}$ is a measurable set and considers the measurable set

$$D = \bigcup_{M \geq 1} D^M$$

with

$$D^M = \bigcup_{m \geq 1} \bigcap_{n \geq m} \bigcup_{0 < i \leq n} \bigcap_{i < j < i+3} D_{i,j}^{M,n}$$

$$= \underline{\lim} \bigcup_{0 < i \leq n} \bigcap_{i < j < 1+3} D_{i,j}^{M,n} \text{ .}$$

D is the event that there exists an integer M such that for all n sufficiently large the previous inequality holds at some point i/n . Hence $\bigcup_{t\in[0,1]} \mathcal{D}_t \subseteq D$. Now

$$P(D^M) \leq \underline{\lim}\; n\left[P(|W_{1/n}|< \frac{7M}{n})\right]^3 = 0 \text{ .}$$

Hence $\qquad P(D) = 0$.

If we have assumed that we work with a complete probability space (see Appendix), $\bigcup_{t\in[0,1)} \mathcal{D}_t$ is measurable with probability zero.

II.2 Stochastic Integration

Let us briefly discuss some basic facts about stochastic integral (see Appendix). The problem is to define an integral of the form

$$I = \int_0^t F(W_\tau)\, dW_\tau \text{ .}$$

The following theorem about property of the path of the Wiener process will make clear the difficulty we encounter by trying to define such an integral.

Theorem (2.4): Let W be a d-dimensional Wiener process and define for $\alpha > 0$, and n positive integer.

$$g(n,\alpha) = \sum_{k=1}^{2^n} | W_{k2^{-n}} - W_{(k-1)^{-n}}|^{\alpha}$$

then with probability one we have

i) if $\alpha < 2$, then $g(n,\alpha) \to \infty$ as $n \to \infty$

ii) if $\alpha = 2$, then $g(n,\alpha) = d$ as $n \to \infty$

iii) if $\alpha > 2$, then $g(n,\alpha) = 0$ as $n \to \infty$

Moreover the convergence in ii) and iii) is valid in any L^P space with $p < \infty$.

Proof: Using the independence of the increments of the Wiener process and the estimate $\mathbb{E}[|x - \mathbb{E}(x)|^2] \leq \mathbb{E}[x^2]$ we obtain

$$\mathbb{E}[|g(n,\alpha) - 2^{n(1-\frac{\alpha}{2})} m_\alpha|] \leq 2^{n(1-\alpha)} m_{2\alpha}$$

where $m_\alpha = \mathbb{E}[|W|^\alpha]$ and in particular $m_2 = d$.

Let $\alpha < 2$ then for any fixed λ, $\frac{1}{2} 2^{n(1-\frac{\alpha}{2})} m_\alpha > \lambda$ for n sufficiently large. For such n

$$\begin{aligned} P(g(n,\alpha) < \lambda) &\leq P((g(n,\alpha) - 2^{n(1-\frac{\alpha}{2})} m_\alpha) < -\frac{1}{2} 2^{n(1-\frac{\alpha}{2})} m_\alpha) \\ &\leq P(|g(n,\alpha) - 2^{n(1-\frac{\alpha}{2})} m_\alpha|) > \frac{1}{2} 2^{n(1-\frac{\alpha}{2})} m_\alpha) \\ &\leq \frac{\mathbb{E}(|g(n,\alpha) - 2^{n(1-\frac{\alpha}{2})} m_\alpha|^2)}{\frac{1}{4} 2^{n(1-\frac{\alpha}{2})} m_\alpha} \end{aligned}$$

by the Doob's inequality

then $$P(g|n,\alpha) < \lambda) \leq 4 \frac{m_{2\alpha}}{m_{\alpha^2}} 2^{-n} .$$

Thus using the first Borel Cantelli lemma $g(n,\alpha) \geq \lambda$ for large n since λ is arbitrary and part i) of the theorem is proved. The proof of ii) and iii) is similar. We have to estimate $P(|g(n,\alpha)| > \varepsilon)$ respectively. The L^P convergence can be proved by direct computation on $\mathbb{E}[|g(n,2) - d|^{2m}]$ in case ii) and by using of the triangular inequality in the last case.

From this theorem it follows that the typical paths of Wiener process are not of bounded variation. On the other side, if we consider for symplicity the case of Brownian motion in dimension one, the Wiener measure is concentrated on continuous function having on each interval [0,1] a quadratic variation given by

$$P_W - \lim \sum_{k=1}^{(t2^n)} | W_{k\, 2^{-n}} - W_{(k-1)2^{-n}}|^2 = \sigma^2 t \ .$$

For each function in this class Itô's formula is valid (see Appendix) i.e. for each C^2-function F we have

$$F(W_t) - F(W_o) = \int_o^t F'(W_s + \frac{\sigma^2}{2} \int_o^r F''(W_s)\, ds \qquad (2.22)$$

when the first integral in the right hand side of the above expression is defined as the limit for $n \to \infty$ of the following expression

$$\sum_{k=1}^{[t2^n]} F'(W_{(k-1)2^{-n}}) [W_{k2^{-n}} - W_{(k-1)2^{-n}}] \qquad (2.23)$$

Itô's formula is the basis of the stochastic calculus based on Wiener's process.

The Brownian motion is a continuous martingale and plays a central role in the class of continuous martingales. Let P_t be the σ-algebra generated by $\{W_s\}_{0<s<t}$. A stochastic process M_t is <u>adapted</u> with respect to <u>filtration</u> $\{P_t\}_{t\in I}$ if M_t is for each t P_t-measurable. An adapted process M_t is called a <u>martingale</u> (P_t-martingale) if

$$\mathbb{E}[M_t - M_s | P_s] = 0 \quad \forall\, s \leq t \ .$$

From the martingale property of the Wiener process and from the special form of the approximation (2.23) it follows that the stochastic integrable occuring in Itô's formula are (local) martingales (see Appendix). Especially the stochastic process defined as

$$W_t^2 - \sigma^2 t = 2 \int_o^t W_s \cdot dW_s$$

is a martingale and this property characterizes the Wiener process in the class of continuous martingales.

Since each typical path of the Wiener process is not of bounded variation it is clearly impossible, H_t being an adapted process, to define the integral $\int_I H_t\, dW_t$ using methods of the usual theory of integration. For the special integrands for which Itô's formula is applicable it is possible to us the pathwise construction discussed above. For more general integrands the L^2-construction of Itô must be used. Setting

$$\int_I H(t)\, dW_t = \sum_i h_i\, [W_{t_i+1} - W_{t_i}] \tag{2.24}$$

for elementary integrand of the form

$$H(t,\omega) = \sum_i h_i(\omega) \quad I_{[t_i,t_{i+1}]}(t) \tag{2.25}$$

$I_{[t_i,t]}$ being the characteristic function of the interval $[t_i,t_j]$ and using the martingale property of the Brownian motion and the independence of the increments we have the following isometry

$$\mathbb{E}\left[\left(\int_I H(t) dW_t\right)^2 \right] = \mathbb{E}\left[\int_I (H(t))^2 dt \right] .$$

Using now this isometry the stochastic integral can be extended to a more general class of integrands (for more detail see e.g. [68]).

II.3 Diffusion Process

The Wiener process is the prototype of a diffusion. If we drop the assumption of homogeneity and isotropy (e.g. if some external force is present), we can no longer expect to have $\mathbb{E}[d\xi_t | \xi_t] = 0$, but rather $\mathbb{E}[d\xi_t | \xi_t] = f(\xi_t, t, dt)$ for some function f, if this expectation exists.

In view of the description given at the beginning (2.1) we will assume that

$$\mathbb{E}[d\xi_t | \xi_t] = b(\xi_t, t) dt + o(dt) . \tag{2.26}$$

This assumption is both intuitively reasonable and technically powerful.

Next, note that for the Wiener process dW_t is independent of all the random variables $W_s, s \leq t$, but we cannot expect this property in the more general situation. Indeed, in the situation discusses above, if $d\xi_t$ is independent of ξ_t, then $\mathbb{E}(d\xi_t | \xi_t) = \mathbb{E}(d\xi_t)$, and the function $b(x,t)$ would have to be spacially constant. We can, however, assume that where the particle moves to between time t and $t+dt$ depends only of the position of the particle at time t and not on any of the rest of the history of the particle. In other words, the past and the future of the particle may not be independent, but given the present, its past and future are independent. This is the Markov property. In the following we always assume that the process $\xi(t)$ verifies this property, exept mentioned otherwise.

More generally, the covariance of the process depends on the position and time, namely for a process in $\mathbb{R}^d$ we can have

$$\mathbb{E}[d\xi_t^i \, d\xi_t^i \mid \xi_t] = \sigma^{ij}(\xi_t,t)dt + o(dt) \tag{2.27}$$

with the conditional moments of higher order still equal to $o(dt)$.

This suggests to develop a model of mechanics based on the Markovian diffusion process $\xi(t)$ in $\mathbb{R}^d$ given by

$$d\xi_t^i = b^i(\xi_t,t)dt + \sum_{j=1}^{d} \sigma^{ij}(\xi_t,t)dW_t^j \tag{2.28}$$

where now W_t^i are d independent standard Brownian motion, $W_o^i = o$

$$\mathbb{E}[dW_t \mid \xi_t = x] = 0$$

$$\mathbb{E}[dW_t^i \, dW_t^j \mid \xi_t = x] = \delta^{ij}dt$$

or, in integral form

$$\xi_t^i(s,x) = x^i + \int_s^t b^i(\xi_\tau(s,x),\tau)d\tau + \sum_{j=1}^{n} \int_s^t \sigma^{ij}\xi_\tau(s,x),\tau)dW_\tau^j, 0\le s\le t \tag{2.29}$$

where the stochastic integral is taken in the sense of Itô [69].

To this process we can associate a semigroup, the generator of which is defined by

$$A_t f(x) = \lim_{h\downarrow 0} \frac{\mathbb{E}_{x,t}[f(\xi_{t+h}(x,t)] - f(x)]}{h} \tag{2.30}$$

on $C_o^\infty(\mathbb{R}^d)$ function and can easily be computed using Itô's formula, which takes the following form for a diffusion process:

Lemma (2.5): (Itô's formula)

Let u be a function defined for $t \in [0,T], x \in \mathbb{R}^d$, continuous with first order derivatives in t and x and second derivatives in x, then

$$du(\xi_t,t) = \left\{\frac{\partial u}{\partial t} + \sum_{i=1}^{d} b^i \frac{\partial u}{\partial x^i} + \frac{1}{2} \sum_{i,j,k=1}^{d} \sigma^{ij}\sigma^{jk} \frac{\partial u}{\partial x^i \partial x^k}\right\} (\xi_t,t)dt$$

$$+ \sum_{i,j=1}^{d} (\sigma^{ij} \frac{\partial u}{\partial x^i}) (\xi_t,t)dW^j(t) . \qquad (2.31)$$

Then A_t is represented on $C_o^\infty(\mathbb{R}^d)$ by

$$A_t = \sum_{i=1}^{d} b^i(x,t) \frac{\partial}{\partial x^i} + \frac{1}{2} \sum_{i,j,k=1}^{d} \sigma^{ij}(x,t)\, \sigma^{jk}(x,t) \frac{\partial^2}{\partial x^i \partial x^k} . \qquad (2.32)$$

This generator is called the forward generator of ξ_t.

Since we are interested in the Markov process, we have as for the Brownian motion a transition probability

$$p(x,t,\mathcal{B},s) = P[\xi_s \in \mathcal{B} \mid \xi_t = x] \qquad (2.33)$$

for $\mathcal{B}$ a Borel set of $\mathbb{R}^d$ and $t \leq s$. This function verifies the backward Kolmogorov equation (i.e. p is regarded as a function of the starting point)

$$\frac{\partial}{\partial t} p(x,t,\mathcal{B},s) = - A_t p(x,t,\mathcal{B},s) . \qquad (2.34)$$

In the case where p is absolutely continuous with respect to the Lebesgue measure

$$p(x,t,dy,s) = p(x,t,y,s)dy$$

the density of probability p verifies the Fokker-Planck equation or forward Kolmogorov equation (differentiation now with respect to the final point)

$$\frac{\partial}{\partial s} p(x,t,y,s) = A_s^* p(x,t,y,s) \qquad (2.35)$$

where A_t^* is the formal L^2-adjoint of A_t, that is

$$\frac{\partial}{\partial s} p(x,t,y,s) = - \sum_{i=1}^{d} \frac{\partial}{\partial y^i} (b^i(y,s)p(x,t,y,s)$$

$$+ \frac{1}{2} \sum_{i,j,k=1}^{d} \frac{\partial^2}{\partial y^i \partial y^i} \sigma^{ik}(y,s)\, \sigma^{kj}(y,s)\, p(x,t,y,s). \qquad (2.36)$$

In the case where $\sigma^{ij}(x,t) = \sigma\delta^{ij}$ with σ constant, this formula reduces to

$$\frac{\partial}{\partial s} p(x,t,y,s) = - \nabla_y \cdot (b(y,s)p(x,t,y,s)) + \frac{\sigma^2}{2} \Delta_y \, p(x,t,y,s) \, . \tag{2.37}$$

II.4 Kinematics of Diffusion Processes

As before, let Ω be the space of continuous functions $\omega: I \to \mathbb{R}^d$ where I can be $\mathbb{R}$, $\mathbb{R}_+$ or a finite interval $[0,T]$ of time, equipped with the topology of uniform convergence on compacts. That is, the topology induced by the family of semi-norms $\varrho_K(f) = \sup_{t\in K}|f(t)|$ for all compact subsets K of I. This makes the trajectory space Ω a complete separable metrizable space (Polish space), which is a Banach space in the case where $I = [0,T]$. Hence we can look at Ω as a measurable space with respect to the Borel field $\mathcal{B}$. Now, suppose we are given a $\mathcal{B}$-probability measure P on Ω, we obtain a probability space $(\Omega,\mathcal{B},P)$. The Borel functions f on Ω become random variables, and we denote their expectation provided they are integrable, as previously by $\mathbb{E}[f] = \int_\Omega f(\omega)\,P(d\omega)$. If Σ is a sub-σ-algebra of $\mathcal{B}$, and if $f \in L^1(\Omega,P)$, then we will denote the conditional expectation of f, given Σ, by $\mathbb{E}[f|\Sigma]$.

At each time t the state of the system we consider is described by a random function $\xi_t : \omega \to \omega(t) = \xi_t(\omega)$, called the configuration process (or the <u>coordinate process</u>) which is at each time a $\mathcal{B}$-measurable function. Hence $\mathcal{B} = \sigma\{\xi_t, t \in I\}$, the σ-algebra is generated by the family $\{\xi_t\}_{t\in I}$. Certain sub-σ-algebras of $\mathcal{B}$ are of special interest.

The <u>past</u> at time t, $\mathcal{P}_t$, for the process ξ_t, is given by $\mathcal{P}_t = \sigma\{\xi_s, s \leq t\}$. The family $\{\mathcal{P}_t\}_{t\in I}$ provides an increasing filtration, called <u>natural</u> or <u>standard filtration</u> of the coordinate process. In view of martingale calculation (see section IV) it is convenient to introduce a new σ-field $\overline{\mathcal{B}} = \sigma\{\mathcal{B},N\}$ where N is the family of all P-null sets, called the P-completion of $\mathcal{B}$, and a new filtration $\overline{\mathcal{P}}_t$ verifying $\overline{\mathcal{P}}_t = \bigcap_{\delta>0} \sigma\{\mathcal{P}_{t+\delta},N\}$. The new family is increasing, right continuous and $\overline{\mathcal{P}}_0(\overline{\mathcal{P}}_{-\infty})$ contains all P-null sets, we refer to these conditions as <u>"the usual conditions"</u>. The intuitive meaning of these filtrations is that a Borel function on Ω is $\mathcal{P}_t(\overline{\mathcal{P}}_t)$-measurable precisely when it only depends on configurations up to time t.

The <u>future</u> at time t, $\mathcal{F}_t$, is given by $\mathcal{F}_t = \sigma\{\xi_u, u \geq t\}$. Hence $\{\mathcal{F}_t\}_{t\in I}$ define a decreasing filtration. If in a similar way as before we introduce the P-completion $\overline{\mathcal{F}}_t$ is a decreasing, left continuous family such that $\overline{F}_t(\overline{F}_{+\infty})$ contains all P-null sets. Finally, the present at time t, N_t, is given by $\sigma\{\xi_t\}$ and we denote by $\overline{N}_t$ its P-completed version.

In the following, we are interested in processes ξ_t which are Markovian and of diffusion type in $\mathbb{R}^d$

$$d\xi_t = b_+(\xi_t,t)dt + \sigma dW_t \tag{2.38}$$

where W_t^i are n-independent standard Wiener processes with covariance $\mathbb{1}t$ and for the sake of simplicity we assume that the diffusion coefficient is a constant (hence the identity in $\mathbb{R}^d$). Moreover, we impose some smoothness condition on the drift $b_+(x,t)$: b_+ is a smooth function bounded by $C(1+|x|)$ for some constant C. The forward drift (or velocity) b_+ gives the best prediction based on information in the past at time t of how the configuration will change just after time t. The forward generator A_t (2.32) for this process takes now the form

$$A_t = b_+(x,t) \cdot \nabla + \frac{\sigma^2}{2} \Delta . \tag{2.39}$$

In order to describe the kinematics of such a process we have to define a reasonable notion of derivatives. From previous properties neither ξ_t nor any function $f(\xi_t)$ is differentiable in the usual sense. Following E.Nelson [90] we introduce a regularization procedure to eliminate too strong irregularities of trajectories. The method consists in the use of conditioning with respect to the past P_t and taking the mean over all possible values of ξ_{t+dt}, $dt > 0$.

Hence we define a forward derivative D_+ in the following way. Let $f(x,t) \in C^2(\mathbb{R}^d \times \mathbb{R})$

$$D_+f(\xi_t,t) = \lim_{h\downarrow 0} \mathbb{E}\left[\frac{f(\xi_{t+h},t+h) - f(\xi_t,t)}{h} \middle| P_t\right]. \tag{2.40}$$

As the considered process is Markovian, the previous definition reduces to

$$D_+f(\xi_t,t) = \lim_{h\downarrow 0} \mathbb{E}\left[\frac{f(\xi_{t+h},t+h) - f(\xi_t,t)}{h} \middle| N_t\right]. \tag{2.41}$$

This implies the following formula

$$D_+f(x,t) = \lim_{h\downarrow 0} \mathbb{E}\left[\frac{f(\xi_{t+h},t+h) - f(x,t)}{h} \middle| \xi_t = x\right]. \tag{2.42}$$

From this definition, taking into account (2.39) it follows

$$D_+\xi_t = b_+(\xi_t,t) \tag{2.43}$$

and using Itô's calculus one obtains the explicit formula for D_+f

$$D_+f(x,t) = \frac{\partial f}{\partial t}(x,t) + b_+(x,t)\circ\nabla f(x,t) + \frac{\sigma^2}{2}\Delta f(x,t) . \tag{2.44}$$

Now let us assume that ξ_t has at each time t a smooth density $\rho(x,t)$ with respect to the Lebesgue measure, that is

$$P[\xi_t \in A] = \int_A \rho(x,t)dx \tag{2.45}$$

for all Borel sets A in $\mathbb{R}^d$.

Hence for all integrable functions f the expectation of f is given by

$$\mathbb{E}[f(\xi_t)] = \int_{\mathbb{R}^d} f(x)\ \rho(x,t)dx . \tag{2.46}$$

Assuming that f is smooth and using the definition of the generator A_t of the process we obtain

$$\begin{aligned}\frac{d}{dt}\mathbb{E}[f(\xi_t)] &= \int_{\mathbb{R}^d} f(x)\ \frac{\partial}{\partial t}\rho(x,t)dx \\ &= \int_{\mathbb{R}^d} (A_t f)(x)\ \rho(x,t)dx \\ &= \int_{\mathbb{R}^d} (b_+\cdot\nabla + \frac{\sigma^2}{2}\Delta)\ f(x,t)\ \rho(x,t)dx\end{aligned}$$

and integrating by part

$$\frac{\partial\rho}{\partial t} = -\nabla\cdot(b_+\rho) + \frac{\sigma^2}{2}\Delta\rho . \tag{2.47}$$

The density ρ verifies (in the weak sense) the Fokker-Planck equation (forward Kolmogorov equation). If b_+ is smooth and if the process has a smooth density at initial time, this property holds for all times in the considered time interval. In chapter IV we will discuss such a condition.

Let us also notice that the following property holds

$$\frac{d}{dt}\mathbb{E}[f(\xi_t,t)] = \mathbb{E}[D_+f(\xi_t,t)] . \tag{2.48}$$

Now let us assume that we are able to define the following operation

$$D_-f(\xi_t,t) = \lim_{h\downarrow 0} \mathbb{E}\left[\frac{f(\xi_t,t)-f(\xi_{t-h},t-h)}{h} \,\middle|\, F_t\right] \tag{2.49}$$

called the <u>backward derivative</u> which we can rewrite for Markov processes

$$D_-f(\xi_t,t) = \lim_{h\downarrow 0} \mathbb{E}\left[\frac{f(\xi_t,t)-f(\xi_{t-h},t-h)}{h} \,\middle|\, N_t\right] \tag{2.50}$$

and

$$D_-f(x,t) = \lim_{h\downarrow 0} \mathbb{E}\left[\frac{f(x,t)-f(\xi_{t-h},t\ h)}{h} \,\middle|\, \xi_t=x\right] \tag{2.51}$$

To derive the explicit form of D_-f, let us establish for smooth functions f and g the following useful derivation formula

$$\frac{d}{dt}\mathbb{E}[f(\xi_t,t)g(\xi_t,t)] = \mathbb{E}[(D_+f)(\xi_t,t)g(\xi_t,t)]+\mathbb{E}[f(\xi_t,t)D_-g(\xi_t,t)] \,. \tag{2.52}$$

Hence for $a < b$ in I we have to prove

$$\mathbb{E}[f(\xi_b,b)g(\xi_b,b) - f(\xi_a,a)g(\xi_a,a)]$$
$$= \int_a^b \mathbb{E}[(D_+f)(\xi_t,t)g(\xi_t,t) + f(\xi_t,t)D_-g(\xi_t,t)]dt \,.$$

Let us assume f and/or g in $C_0^2(\mathbb{R}^d xI)$ and define $F(t)=f(\xi_t,t)$, $G(t)=g(\xi_t,t)$ and make a partition of the time interval (a,b)

$$t_j = a + \frac{j(b-a)}{n} \qquad j = 0,1,\ldots,n$$

then

$$\mathbb{E}[F(b)G(b) - F(a)G(a)] = \lim_{n\to\infty}\sum_{j=1}^{d-1} \mathbb{E}[F(t_{j+1})G(t_j) - F(t_j)G(t_{j-1})]$$

$$= \lim_{n\uparrow\infty}\sum_{j=1}^{n-1}\left[(F(t_{j-1})-F(t_j))\frac{G(t_j)-G(t_{j-1})}{2} + \frac{F(t_{j+1})+F(t_j)}{2}(G(t_j)-G(t_{j-1}))\right]$$

$$= \lim_{n\uparrow\infty}\sum_{j=1}^{n-1} \mathbb{E}[(D_+F)(t_j)G(t_j) + F(t_j)D_-G(t_j)]\ \frac{b-a}{n}$$

$$= \int_o^b \mathbb{E}[(D_+F)(t)G(t) + F(t)D_-G(t)]dt \,.$$

Notice that relation (2.52) gives an integration by part formula. Moreover, the following derivation formula holds

$$\frac{d}{dt}\,\mathbb{E}[f(\xi_t,t)] = \mathbb{E}[D_\pm f(\xi_t,t)] \;. \tag{2.53}$$

Now, for f and g in $C_o^2(\mathbb{R}^d \times I)$ we have

$$\frac{d}{dt}\int_I \mathbb{E}[f(\xi_t,t)g(\xi_t,t)]dt = 0$$

hence by formula (2.52)

$$0 = \int_I \mathbb{E}[(D_+f)(\xi_t,t)g(\xi_t,t) + f(\xi_t,t)D_-g(\xi_t,t)]dt$$

then

$$\int_I\int_{\mathbb{R}^d}[(\frac{\partial}{\partial t} + b_+\cdot\nabla + \frac{\sigma^2}{2}\Delta)f](x,t)g(x,t)\rho(x,t)\;dx\;dt$$

$$= -\int_I\int_{\mathbb{R}^d} f(x,t)(D_-g)(x,t)\rho(x,t)\;dx\;dt\;.$$

Integrating by part the left hand side we obtain

$$\int_I\int_{\mathbb{R}^d} f(x,t)[(-\frac{\partial}{\partial t} - b_+\circ\nabla - \nabla\circ b_+ + \frac{\sigma^2}{2}\Delta)g\rho](x,t)\;dx\;dt$$

$$= -\int_I\int_{\mathbb{R}^d} f(x,t)(D_-g)(x,t)\;\rho(x,t)\;dx\;dt\;.$$

Hence

$$\int_I\int_{\mathbb{R}^d} f(x,t)[(-\frac{\partial}{\partial t} - b_+\cdot\nabla + \sigma^2\frac{\nabla\rho}{\rho}\cdot\nabla + \frac{\sigma^2}{2}\Delta)g](x,t)\;\rho(x,t)\;dx\;dt$$

$$+\int_I\int_{\mathbb{R}^d} f(x,t)(D_-g)(x,t)\rho(x,t)dx\;dt = -\int_I\int_{\mathbb{R}^d} f(x,t)(D_+g)(x,t)\rho(x,t)dx\;dt$$

using the Fokker-Planck equation (2.47), the second integral of the left hand side vanishes, since f and g are in $C_o^2(\mathbb{R}^n \times I)$ then if we define b_- by

$$b_-(x,t) = b_+(x,t) - \sigma^2\,\frac{\Delta\rho(x,t)}{\rho(x,t)} \tag{2.54}$$

we obtain

$$D_-g(x,t) = \frac{\partial g}{\partial t}(x,t) + b_-(x,t)\cdot\nabla g(x,t) - \frac{\sigma^2}{2}\Delta g(x,t)\;. \tag{2.55}$$

In particular, for processes for which (2.54) is meaningful

$$D_-\xi_t = b_-(\xi_t,t)$$

$b_-(x,t)$ is called the backward drift (velocity).

The physical interpretation of the forward and backward drift is the following: $b_+(x,t)$ is the mean velocity of particles leaving x at time t and $b_-(x,t)$ is the mean velocity of particles entering x at time t.

More generally, in the following we will say that under the probability P the configuration process ξ_t is a (Nelson) smooth diffusion with coefficient $\sigma^{ij}(x,t)$, $b_+^i(x,t)$ and $b_-^i(x,t)$, $1 \leq i, j \leq n$ in case

i) the configuration process is Markovian;

ii) the σ^{ij} are all smooth bounded functions where the $\sigma^{ij}(x,t)$ are entries of a positive definite matrix. The b_+^i and b_-^i are all smooth functions bounded by $C(1+|x|)$ for some constant C.

iii) The following limits exist for any $f \in C_0^\infty(\mathbb{R}^d)$

$$\lim_{h\downarrow 0_+} \mathbb{E}\left[\frac{f(\xi_{t+h}) - f(\xi_t)}{h}\,\middle|\, N_t\right] = \sum_{ij}\left[\frac{1}{2}\sigma^{ij}\frac{\partial^2}{\partial x^i \partial x^j} + b_+^i \frac{\partial}{\partial x}\right] f(\xi_t)$$

$$\lim_{h\downarrow 0_+} \mathbb{E}\left[\frac{f(\xi_t) - f(\xi_{t-h})}{h}\,\middle|\, N_t\right] = \sum_{ij}\left[-\frac{1}{2}\sigma^{ij}\frac{\partial^2}{\partial x^i \partial x^j} + b_-^i \frac{\partial}{\partial x}\right] f(\xi_t).$$

Hence we have two kinds of velocity (forward and backward). This suggests that for the processes we consider we can define a notion of time reversal, we will look at this problem in the next section.

Let us remark that the Fokker-Planck equation can be reformulated in terms of the backward drift b_-,

$$\frac{\partial \rho}{\partial t} = -\nabla \cdot (b_-\rho) - \frac{\sigma^2}{2}\Delta\rho . \tag{2.56}$$

It is convenient to define other "time-derivations"

$$D = \frac{1}{2}(D_+ + D_-) \tag{2.57}$$

and

$$\delta D = \frac{1}{2}(D_+ - D_-) \tag{2.58}$$

so that

$$D_{\pm} = D \pm \delta D \ . \tag{2.59}$$

We introduce two other velocities associated with D and δD

$$v(x,t) = \frac{1}{2} (b_+(x,t) + b_-(x,t)) \tag{2.60}$$

the <u>current velocity,</u> and

$$u(x,t) = \frac{1}{2} (b_+(x,t) - b_-(x,t)) \tag{2.61}$$

the <u>osmotic velocity.</u> We emphasize that

$$u(x,t) = \frac{\sigma^2}{2} \frac{\nabla \rho(x,t)}{\rho(x,t)} = \frac{\sigma^2}{2} \nabla \log \rho(x,t) \tag{2.62}$$

u is always a gradient.
Hence

$$b_{\pm}(x,t) = v(x,t) \pm u(x,t) \tag{2.63}$$

and the action of D and δD on smooth function f have an expression in terms of v and u

$$Df(x,t) = \frac{\partial f}{\partial t} (x,t) + v(x,t) \cdot \nabla f(x,t) \tag{2.64}$$

and

$$\delta Df(x,t) = u(x,t) \cdot \nabla f(x,t) + \frac{\sigma^2}{2} \Delta f(x,t) . \tag{2.65}$$

Notice that D is a derivation (in the usual sense) which is not the case for $D_{\pm}$ and δD . D can be seen as the derivative along the flow defined by the field $v(x,t)$. Moreover, the Fokker-Planck equations (2.47) and (2.57) can now be put in the form of a <u>continuity equation</u>

$$\frac{\partial \rho}{\partial t} + \nabla \cdot (\rho v) = 0 \tag{2.66}$$

and of the <u>osmotic equation</u>

$$\frac{\sigma^2}{2} \Delta \rho = \nabla \cdot (\rho u) \ . \tag{2.67}$$

Equation (2.66), which can be directly compared with the hydrodynamical conservation law (1.40), (1.50), is strictly connected with the continuity in time of the random trajectories.

Finally, let us remark that for smooth density the time derivative of the osmotic velocity exists and the continuity equation (2.66) and the osmotic equation (2.67) implies

$$\frac{\partial u}{\partial t} = -\frac{\sigma^2}{2}\nabla(\nabla\cdot v) - \nabla(u\cdot v) \ . \tag{2.68}$$

II.5 The Time-Reversed Diffusion Process

II.5a Brownian Motion with Lebesgue Measure as Initial Distribution

As we have seen before, the Wiener process is a Markov process, hence its probabilistic behaviour in the future is completely determined by its state at the present time. This behaviour is described by the transition probability density

$$p(y,s,x,t) = \frac{1}{(2\pi(t-s)\sigma^2)^{d/2}}\, e^{-\frac{(x-y)^2}{2\sigma^2(t-s)}} = \underline{p}(x-y,t-s) \tag{2.69}$$

for $t > s$, and σ^2 a positive constant; with $\underline{p}$ defined in (2.8).

The distribution of the Wiener process (with diffusion coefficient σ^2) is determined by the transition probability and its initial distribution. Let us know consider the case where the initial distribution is the Lebesgue measure on $\mathbb{R}^d$, hence the initial distribution is not a probability measure! Nevertheless, the Lebesgue measure is an invariant measure for the Wiener process, that is, if at an initial time the process is given with a Lebesgue distribution, the same occurs for any time.

Because of the Lebesgue measure we have not an underlying probability space but we can introduce a measure space $(\Omega,\underline{F},\mu)$ with a σ-finite measure μ and define the notion of martingale. Let $\{\underline{\underline{F}}_t\}_{t\in[0,T]}$, $\underline{\underline{F}}_t \subset \underline{\underline{F}}$ an increasing family of sub-σ-algebras.

A map $(t,\omega) \to X_t(\omega) : \Omega \times [0,T] \to \mathbb{R}^d$ is called a <u>martingale</u> (with respect to the family $\underline{\underline{F}}_t$ and the σ-finite measure μ) if

i) X_t is $\underline{\underline{F}}_t$-measurable $\forall\, t \in [0,T]$.

ii) For $t > s$, we have $\int_A X_s d\mu = \int_A X_t d\mu \ \forall\, A \subset \underline{\underline{F}}_s$ such that $\int_A |X_t| d\mu < \infty$.

Hence the Wiener process W_t is defined on the measure space $(\Omega,\mathcal{B},\mu)$ where Ω is the space of continuous trajectories, $\mathcal{B}$ the

Borel σ-algebra and μ the measure induced by the initial Lebesgue distribution and the transition probability (2.69). If we denote by $P_t^{[W]}$ the natural filtration of the Wiener process W_t , then W_t is a $P_t^{[W]}$-martingale in the above sense.

Now, let us consider the time reversal process $\check{W}_t$ defined by

$$\check{W}_t = W_{T-t} \tag{2.70}$$

where W_t is the Wiener process with initial Lebesgue distribution.

The time-reversal process $\check{W}(t)$ is also a Markov process adapted to the filtration $P_t^{[\check{W}]} = F_{T-t}^{[W]}$, $t\in[0,T]$. For all continuous functions $f(x)$, $g(x)$ with compact support in $\mathbb{R}^d$ we have the following

$$\begin{aligned}\int_\Omega f(W_s)g(W_t)d\mu &= \int_{\mathbb{R}^d}\int_{\mathbb{R}^d} f(x)\ p(x,s,y,t)g(y)dx\ dy \\ &= \frac{1}{(2\pi(t-s)\sigma^2)^{d/2}}\int_{\mathbb{R}^d}\int_{\mathbb{R}^d} f(x)e^{-\frac{(y-x)^2}{2\sigma^2(t-s)}}g(y)dx\ dy\end{aligned} \tag{2.71}$$

but

$$\int_\Omega f(W_s)g(W_t)d\mu = \int_\Omega f(\check{W}_{t-s})g(\check{W}_{T-t})d\mu\ . \tag{2.72}$$

Since $\check{W}_t$ is also Markov process, from (2.71) and (2.72) we have

$$\begin{aligned}&\int_\Omega f(\check{W}_{T-s})g(\check{W}_{T-t})d\mu = \\ &\frac{1}{\{2\sigma^2[(T-s)-(T-t)]\}^{d/2}}\int_{\mathbb{R}^d}\int_{\mathbb{R}^d} f(x)\ e^{-\frac{|x-y|^2}{2\sigma^2[(T-s)-(T-t)]}}g(y)dx\ dy\ .\end{aligned} \tag{2.73}$$

Consequently, the transition probability density of $\check{W}_t$ is also given by (2.69) and $\check{W}_t$ have the same distribution as W_t , moreover, $\check{W}_t$ is a $P_t^{[\check{W}]}$-martingale.

In particular, the time reversed Wiener process is a Wiener process, when the initial distribution is a Lebesgue measure. Let us remark that in general the time reversed Wiener process is not a Wiener process.

II.5b Time-Reversed Diffusion Process

As previously let Ω be the space of all continuous functions from $[0,t]$ to $\mathbb{R}^d$. Suppose that under probability $P \ll \mu$ (P absolutely continuous probability measure with respect to the measure μ of the Wiener process with initial Lebesgue measure) the configuration process ξ_t is a Markov diffusion process of the following type

$$d\xi_t = b_+(\xi_t,t)\,dt + \sigma dW_t \tag{2.74}$$

where $b_+(\cdot,t)$ is a vector field, sufficiently regular. We are going to prove that $\hat{\xi}_t = \xi_{T-t}$ is also a Markov diffusion process of the previous type. For sake of simplicity we choose $\sigma = \mathbb{1}$.

The following consideration is based on the Girsanov [56] - Cameron-Martin [21]-formula. Let us give a simplified version corresponding to the case where the Wiener process has a Lebesgue initial distribution

Theorem (2.6):

Let μ be a measure on the filtration space $(\Omega,\underline{\underline{F}}_t,\underline{\underline{F}})$ such that under μ, $\xi_t(\omega)$, $t \in [0,T]$ is a Wiener process with Lebesgue's measure dx as initial distribution, and let $\rho(x,0)$ be a non-negative function such that $\int \rho(x,0)dx = 1$. If

$$Q_t = \rho(x,0)\, e^{[\int_0^t b_+(\xi_s,s)d\xi_s - \frac{1}{2}\int_0^t |b_+(\xi_s,s)|^2 ds]}$$

is a martingale, then under the probability P defined by $dP = Q_T d\mu$ the process

$$W_t = \xi_t - \int_0^t b_+(\xi_s,s)ds$$

is also a Wiener process with the same diffusion coefficient as ξ_t and with initial distribution $\rho(x,0)dx$.

Now, let us suppose that $\Omega = C(0,T)$ is the space of continuous real function on $[0,T]$ and $\{\underline{\underline{F}}_t^{[\xi_t]}\},t\in[0,T]$ the natural filtration of the configuration process ξ_t satisfying (2.74) under the measure P such that $dP = Q_T d\mu$ as defined in the previous theorem. Then,

$$\log Q_T = \log \rho(\xi_0,0) + \int_0^T b_+(\xi_s,s)d\xi_s - \frac{1}{2}\int_0^T |b_+(\xi_s,s)|^2 ds \tag{2.75}$$

by Itô's formula

$$\log \rho(\xi_0,0) = \log \rho(\xi_T,T) - \int_0^T \nabla \log \rho(\xi_s,s)d\xi_s - \int_0^T \left(\frac{\partial}{\partial s} + \frac{1}{2}\Delta\right) \log \rho(\xi_s,s)ds$$

using the relation

$$b_+ = b_- + \frac{\nabla\rho}{\rho}$$

and the Fokker-Planck equation in backward form (2.56) we obtain

$$\log Q_T = \log \rho(\xi_T,T) + \int_0^T b_-(\xi_s,s)d\xi_s - \frac{1}{2}\int_0^T |b_-(\xi_s,s)|^2 ds + \int_0^T \nabla\cdot b_-(\xi_s,s)ds \ . \tag{2.76}$$

But the second and last terms of the right hand side combined together become a backward stochastic integral (see Appendix), then if we denote $\check{\xi} = \xi_{T-t}$ we can rewrite

$$\log Q_T = \log \rho(\check{\xi}_0,T) - \int_0^T b_-(\check{\xi}_s,T-s)d\check{\xi}_s - \frac{1}{2}\int_0^T |b_-(\check{\xi}_s,T-s)|^2 ds \ . \tag{2.77}$$

Then define

$$\check{Q}_t = \rho(\check{\xi}_0,T)\ e^{-\left[\int_0^t b_-(\check{\xi}_s,T-s)d\check{\xi}_s + \frac{1}{2}\int_0^t |b_-(\check{\xi}_s,T-s)|^2 ds\right]} \tag{2.78}$$

and

$$d\check{P} = \check{Q}_T\ d\mu \ . \tag{2.79}$$

By the previous theorem $\check{\xi}_t$ is a Markovian diffusion process under $\check{P}$

$$d\check{\xi}_t = -\ b_-(\check{\xi}_t,T-t)dt + dB_t \tag{2.80}$$

where B_t is a Wiener process adapted to the filtration $\underline{\underline{F}}_t^{[\check{\xi}]}$. But from (2.78) and (2.79) we have $\check{P} = P$. Hence under P the $\check{\xi}_t$ is a diffusion process, the time reversed process.

II.6 Stochastic Acceleration

To introduce a notion of acceleration in stochastic kinematics we have to define a second order time derivative. A priori there exist four basic candidates: $D_+D_+\xi$, $D_+D_-\xi$, $D_-D_+\xi$ and $D_-D_-\xi$. If ξ is twice differentiable as a function of time each of them reduce to $d^2\xi/dt^2$. Hence the most general acceleration which reduces to the usual acceleration on smooth functions can be written as a four parameters family

$$a_{\kappa\lambda\mu\nu} = [\kappa D_+D_+ + \lambda D_+D_- + \mu D_-D_+ + \nu D_-D_-]\,\xi \qquad (2.81)$$

with the constraint

$$\kappa + \zeta + \mu + \nu = 1 \;.$$

Under time reversal $\xi(t)$ becomes $\check{\xi}_t = \xi_{T-t}$ and as we have seen before, from (2.80) follows

$$D_+\check{\xi} = -\, b_-\,(\check{\xi}_t, T-t)$$

$$= -\, b_-\,(\xi_{T-t}, T-t)$$

and more generally

$$D_\pm\check{\xi}_t = -\, D_\pm\, \xi_{T-t} \;. \qquad (2.82)$$

If we ask for invariance under time reversal, the family of accelera tions reduces to a one parameter family

$$a_\kappa = a_{\kappa,\, \frac{1}{2}-\kappa,\, \frac{1}{2}-\kappa,\, \kappa}$$

$$a_\kappa = [(D_+D_+ + D_-D_-) + (1/2 - \kappa)(D_+D_- + D_-D_+)]\xi_t \qquad (2.83)$$

which can be written as

$$a_\kappa = \left[\left(\frac{D_+ + D_-}{2}\right)^2 + \beta\left(\frac{D_+ - D_-}{2}\right)^2\right]\xi_t,\quad \beta = 4\kappa - 1 \qquad (2.84)$$

$$a_\kappa = \left(\frac{\partial v}{\partial t} + (v\cdot D)v\right) + \beta((u\cdot\nabla)u + \frac{\sigma^2}{2}\Delta u)\,. \qquad (2.85)$$

The expression (2.85) is very remainding of the Madelung fluid equation. Indeed setting $v = \frac{\nabla S}{m}$ in (1.52) and taking the gradient the two equations coincide if we choose $\sigma = \left(\frac{\hbar}{m}\right)^{1/2}$ and $\beta = -1$. Then this consideration is a motivation for the choice $\beta = -1$ and the stochastic acceleration takes the form

$$a = \frac{1}{2}\,(D_+D_- + D_-D_+)\,\xi \;. \qquad (2.86)$$

Let us now consider the following Markov process in one dimension

$$d\xi_t = - \omega\xi_t dt + \sigma dW_t \tag{2.87}$$

where W_t is a Wiener process, and with the invariant Gaussian measure as initial distribution

$$\rho_o(x) = \frac{1}{\sqrt{\omega\pi\sigma^2}} e^{-\frac{\omega x^2}{\sigma^2}} \tag{2.88}$$

then

$$D_+\xi_t = - \omega\xi_t \tag{2.89}$$

$$D_-\xi_t = \omega\xi \tag{2.90}$$

Hence

$$v(x) = 0 \quad \text{and} \quad u(x) = - \omega x \tag{2.91}$$

with the choice (2.86) for the acceleration

$$a = - \omega^2\xi_t \tag{2.92}$$

which is just the deterministic Newton law for the harmonic oscillator.

This process is naturally anociated with the fundamental state of the quantum harmonic oscillator when $\sigma^2 = \frac{\hbar}{m}$, as we will see in section III.3. For other motivations using variational principles see e.g. [62,b] [90,a,e] [112] and chapter V, see also [30,a] for a discussion of the possible choice of β and σ .

Then, according to the previous consideration, we define following E. Nelson the stochastic acceleration by the symmetric expression

$$a = \frac{1}{2} (D_+D_- + D_-D_+)\xi \tag{2.93}$$

using (2.57) - (2.58)

$$a = [D^2 - (\partial D)^2]\xi \tag{2.94}$$

and by (2.60)

$$\begin{aligned} a &= \frac{\partial v}{\partial t} + (v\cdot\nabla)v - (u\cdot\nabla)u - \frac{\sigma^2}{2} \Delta u \\ &= \frac{\partial v}{\partial t} + (v\cdot\nabla)v - \nabla(\frac{1}{2} u^2 - \frac{\sigma^2}{2} \nabla\cdot u) \end{aligned} \tag{2.95}$$

Remark

By trying to generalize stochastic mechanics Davidson [3o a] showed that the diffusion constant ν is not uniquely determined, but rather can be any strictly positive constant of dimension $(\text{length})^2$ divided by time.

The specific value of ν is of course related to a choice of the stochastic acceleration. As discussed above the general form of the stochastic acceleration is

$$a = \varkappa\, D_+D_+ + \lambda\, D_-D_- + \mu\, D_+D_- + \nu\, D_-D_+$$

and the assumption of time reversal invariance yields $\varkappa = \lambda$ and $\mu = \nu$.

Assumming moreover that the coefficents satisfy $\varkappa + \lambda + \mu + \nu = 1$ then the stochastic acceleration reduces to the family of acceleration considered by Davidson namely

$$a_\varkappa = a_D = \left[\left(\frac{D_+ + D_-}{2}\right)^2 + \beta \left(\frac{D_+ - D_-}{2}\right)\right]$$

where $\beta = 4\varkappa - 1$. (Nelson's choice in [90 b] was $\beta = -1$). Davidson then shows that the stochastic Newton law with this acceleration leads to an equation which is equivalent to the Schrödinger equation provided the following choice for β

$$\beta = -\left(\frac{\hbar}{2\, m\nu}\right)^2$$

Let X_D be the diffusion process corresponding to the Davidson model. As a side remark we point out that Ehrenfest's theorem

$$\frac{d^2}{dt^2}\, E\,[X_D] = E\,[a_D]$$

follows from the choice of a_D. In other words Ehrenfest's theorem is just a consequence of time reversal invariance. The particular form of the dynamics and the way how it enters in the theory are of no significance in this respect. Moreover it is easy to show that one cannot use Heisenberg's uncertainty relations in order to single out a specific choice of the stochastic acceleration (see VI. 4). However the Nelson's choice leads to the simplest form for the uncertainty relation [58 a.c.].

II.7 Some Basic Examples

II.7a The Wiener Process

Let us consider the Wiener process in $\mathbb{R}^n$ starting at x_o at time t_o. In this case, the initial distribution is the Dirac measure at x_o, $\mu_o(x) = \delta(x-x_o)$. The distribution at time t admits a probability density $\rho(x,t)$ with respect to the Lebesgue measure

$$\rho(x,t) = \frac{1}{(2\pi\sigma^2(t-t_o))^{n/2}} e^{-\frac{(x-x_o)^2}{2\sigma^2(t-t_o)}} \tag{2.96}$$

where σ^2 is the variance of the process.

Hence $\nabla\log\rho(x,t) = -\frac{1}{\sigma^2}\frac{x-x_o}{t-t_o}$ and the osmotic velocity takes the form

$$u(x,t) = -\frac{1}{2}\frac{x-x_o}{t-t_o}. \tag{2.97}$$

As the forward drift $b_+ = 0$ we have

$$b_+(x,t) = 0 , \quad b_-(x,t) = \frac{x-x_o}{t-t_o} \tag{2.98}$$

and the current velocity assumes the form

$$v(x,t) = \frac{1}{2}\frac{x-x_o}{t-t_o}. \tag{2.99}$$

The stochastic acceleration (2.94) reduces to

$$a(x,t) = \frac{1}{2} D_+b_- = -\frac{1}{2}\frac{x-x_o}{(t-t_o)^2}. \tag{2.100}$$

We can interpret this formula heuristically in the following way. To "prepare" a Brownian motion (starting) at point x_o, we have to apply on the "free" Brownian particle a very strong force and gradually to release it.

Let us remark that the Brownian motion with initial Lebesgue measure is "free" in the sense that the acceleration vanishes.

II.7.b The Brownian Bridge

Let W_t be a one dimensional Wiener process on $[0,T]$ with initial condition $W_o = 0$ and variance $\sigma^2 t$. The process

$$B_t = \frac{T-t}{T} a + W_t + \frac{t}{T}(b-W_T) , \quad t \in [0,T] \tag{2.101}$$

is a Gaussian process such that

$$B_o = \alpha , \quad B_T = \beta \tag{2.102}$$

with mean

$$m(t) = \mathbb{E}[B_t] = \frac{T-t}{T}\alpha + \frac{t}{T}\beta \tag{2.103}$$

and covariance

$$r(s,t) = \mathbb{E}[(B_t - m(t))(B_s - m(s))] = \sigma^2 s \frac{T-t}{T}; \; s \leq t. \tag{2.104}$$

Then the distribution at time t is given by the density

$$\rho^{\sigma}_{\alpha\beta}(x,t) = \sqrt{\frac{T}{2\pi\sigma^2 t(T-t)}} \, e^{-\frac{1}{2\sigma^2}\frac{T}{t(T-t)}(x-m(t))^2} . \tag{2.105}$$

The process B_t is the so-called <u>Brownian Bridge.</u>

Now, let us normalize the process B_t to have zero mean and unit variance. Since the variance of B_t is $\sigma^2 \frac{t(T-t)}{T}$

$$\tilde{B}_t = \sqrt{\frac{T}{\sigma^2 t(T-t)}} \left[B_t - \frac{T-t}{T}\alpha - \frac{t}{T}\beta\right] . \tag{2.106}$$

Then $\tilde{B}_t$ is a Gaussian process with mean zero and covariance

$$\tilde{r}(s,t) = \mathbb{E}[\tilde{B}_t\tilde{B}_s] = \frac{s(T-t)}{t(T-s)} ; \; s \leq t . \tag{2.107}$$

Notice that the distribution of $\tilde{B}_t$ is given by

$$\tilde{\rho}(x,t) = \rho^1_{oo}(x,0) = \frac{1}{\sqrt{2\pi}} e^{-\frac{x^2}{2}} . \tag{2.108}$$

Now, using Hida's representation [65] there exists a standard Wiener process W_t such that

$$\tilde{B}_t = \sqrt{\frac{T(T-t)}{t}} \int_o^t \frac{dW_s}{T-s} . \tag{2.109}$$

Then the process B_t takes the form

$$B_t = \frac{T-t}{T}\alpha + \frac{t}{T}\beta + (T-t)\int_o^t \frac{dW_s}{T-s} \tag{2.110}$$

or, in differential form,

$$dB_t = \left(\frac{\beta-\alpha}{T} - \int^t \frac{dW_s}{T-s}\right)dt + dW_t$$

which can be rewritten also as

$$B_t = \alpha + \int_o^t \frac{\beta-B_s}{T-s}\, ds + W_t \ . \tag{2.111}$$

From (2.111) we obtain for the forward velocity

$$b_+(x,t) = \frac{\beta-x}{T-s} \tag{2.112}$$

and from (2.105) follows for the osmotic velocity

$$u(x,t) = \frac{1}{2}\left(\frac{\alpha}{t} + \frac{\beta}{T-t} - \frac{T}{t(T-t)}\, x\right) \ . \tag{2.113}$$

Hence the backward velocity is given by

$$b_-(x,t) = \frac{x-\alpha}{t} \ . \tag{2.114}$$

We have all the elements to compute the stochastic acceleration obtaining

$$a(x,t) = \frac{\beta-x}{(T-t)^2} + \frac{x-\alpha}{t^2} \ . \tag{2.115}$$

This result extends immediately to the case of an n-dimensional Brownian bridge, if we define it as the n-tuple of n independent one-dimensional Brownian bridges.

II.7c The Bessel Process

Let W_t be a n-dimensional Wiener process of covariance $\sigma^2 \mathbb{1}$. Define

$$R_t = |W_t| = \left(\sum_{i=1}^n W_i^2\right)^{1/2} \tag{2.116}$$

as the radial part of W_t ; by Itô's formula (2.27) we obtain

$$dR_t = \frac{\sigma^2}{2}(n-1)\frac{dt}{R_t} + \sum_{i=1}^n \frac{W_t^i\, dW_t^i}{R_t} \ . \tag{2.117}$$

Using now the fact that a stochastic Itô integral of the form

$y_t(\omega) = \int_0^t a(\tau,\omega)\,dW_\tau(\omega)$, where $a(\tau,\omega)$ is a $m \times n$ matrix, is a m-dimensional Brownian motion if and only if $MM^t = 1$, where M^t denotes the transposed matrix of M (see e.g. [82,b] [114] , it follows that

$$W'_t = \int_0^t \sum_{i=1}^n \frac{W_s^i\, dW_s^i}{R_t} \tag{2.118}$$

is a one dimensional Wiener process with variance σ^2 . Then (2.117) rewrites

$$dR_t = \frac{2}{2}(n-1)\frac{dt}{R_t} + dW'_t \tag{2.119}$$

which is a diffusion equation. The associated generator is given by

$$A = \frac{1}{2}\left(\sigma^2 \frac{\partial^2}{\partial r^2} + \frac{n-1}{r}\frac{\partial}{\partial r}\right) \tag{2.120}$$

where $r = \sqrt{x_1^2 + \ldots + x_n^2}$, it is therefore natural to call this process the Bessel process.

The transition function of the n-dimensional Bessel process is given by

$$p(r_o,t_o,r,t) = \frac{2C_n}{(2\pi\sigma^2(t-t_o))^{n/2}}\, e^{-\frac{(r-r_o)^2}{2\sigma^2(t-t_o)}} \tag{2.121}$$

where the C_n's depend on the parity of n

$$C_{2p+1} = \frac{2p}{1.3.\ldots.(2p+1)}$$

$$C_{2p} = \frac{2}{(p+1)!}\ .$$

Notice that this process has the measure $d\mu(r) = r^{n-1}dr$ as invariant measure. In this particular case we can compute easily the kinematical quantities

$$\begin{aligned} b_+(r,t) &= \sigma^2\,\frac{n-1}{2}\,\frac{1}{r} \\ u(r,t) &= \sigma^2\,\frac{n-1}{2}\,\frac{1}{r} \\ b_-(r,t) &= -\,\sigma^2\,\frac{n-1}{2}\,\frac{1}{r} \\ v(r,t) &= 0 \end{aligned} \tag{2.122}$$

and the stochastic acceleration is given by

$$a(r) = -\sigma^2 \frac{n-3}{r^3} \quad . \tag{2.123}$$

Notice that in the case $n = 3$ we obtain an example of a process which is "free" in the sense that the stochastic acceleration vanishes.

If we suppose that the process has initial distribution given by the Dirac measures at r_o, then the distribution at time $t > t_o$ is

$$d\mu_t(r) = \frac{2C_n}{(2\pi\sigma^2(t-t_o))^{n/2}} \, e^{-\frac{(r-r_o)^2}{2\sigma^2(t-t_o)}} \, r^{n-1} \, dr \quad . \tag{2.124}$$

Notice that the probability that the particle reaches the origin at any time $t > 0$ vanishes.

In the following computation, let us assume for simplicity that $t_o = 0$, and $r_o = 0$.

$$\log \rho(x,t) = \gamma_n - \frac{r^2}{2\sigma^2 t} + (n-1) \log 2$$

where γ_n is a constant with respect to r . Then

$$u(r,t) = -\frac{r}{2t} + \sigma^2 \frac{(n-1)}{2} \frac{1}{r} \tag{2.125}$$

$$b_-(r,t) = \left(\frac{r}{t} - \sigma^2 \frac{n-1}{2} \frac{1}{r}\right) \tag{2.126}$$

and

$$a(r,t) = -\frac{1}{2} \frac{r}{t^2} - \sigma^2 \frac{n-3}{r^3} \quad . \tag{2.127}$$

For a more detailed study of the Bessel process see [82,b].

II.7d The Ornstein-Uhlenbeck Process [106]

Let us, as a last example, consider the Ornstein-Uhlenbeck process described in (1.6) in the case where the configuration space in $\mathbb{R}$.

$$\begin{aligned} dx_t &= v_t \, dt \\ dv_t &= -\beta v_t dt + \beta\sigma \, dW_t \end{aligned} \tag{2.128}$$

where β is a constant and W_t is a standard Brownian motion starting from the origin. Considered as a process on the phase space $\mathbb{R} \times \mathbb{R}$ it is a Markovian diffusion process.

For initial conditions x_0 , v_0 the stochastic differential equation (2.128) rewrites in integral form

$$x_t = x_0 + \int_0^t v_s ds$$
$$v_t = e^{-\beta t} v_0 + \beta\sigma \int_0^t e^{-\beta(t-s)} dw_s \qquad (2.129)$$

the solution v_t of the second equation in (2.18) (the Ornstein Uhlenbeck velocity process) with initial condition v_0 is a Markov-Gaussian process with mean and covariance given by

$$m(t) = \mathbb{E}(v_t) = e^{-\beta t} v_0 \qquad (2.130)$$

$$r(t,s) = \mathbb{E}[(v_t - m(t))(v_s - m(s))] = \frac{\sigma^2}{2\beta}[e^{-\beta|t-s|} - e^{-\beta(t+s)}] \qquad (2.131)$$

the generator of the process can be written

$$A = -\beta v \frac{d}{dv} + \beta^2 \frac{\sigma^2}{2} \frac{d^2}{dv^2} \qquad (2.132)$$

and the transition probability density verifies the Fokker Planck equation

$$\frac{\partial}{\partial t} p(v',t',v,t) = \frac{\partial}{\partial v}(vp(v',t',v,t)) + \beta^2\sigma^2 \frac{\partial^2}{\partial v^2} p(v',t',v,t)$$

which can be easily solved

$$p(v_0,t_0,v,t) = \frac{1}{[\pi\beta\sigma^2(1-e^{-2\beta(t-t_0)})]^{1/2}} e^{-\frac{[v-v_0 e^{-\beta(t-t_0)}]^2}{\beta\sigma^2(1-e^{-2\beta(t-t_0)})}} \qquad (2.134)$$

Notice that the Ornstein Uhlenbeck velocity process with invariant probability distribution

$$d\mu = p(v)\, dv = \frac{1}{\sqrt{\pi\beta\sigma^2}} e^{-\frac{v^2}{\beta\sigma^2}} dv \qquad (2.135)$$

coincides with the oscillator process defined in (II.6.). This process has the tendancy to go toward the origin. Indeed

$$E[v_t] = 0 \tag{2.136}$$

$$E[v_t \, v_{t'}] = \iint vv'p(v',t',v,t)\rho(v')dv'dv$$

$$= \beta\sigma^2 e^{-\beta(t-t')} \quad . \tag{2.137}$$

More generally for arbitrary t and t'

$$\mathbb{E}\,(v_t \, v_{t'}) = \beta\,\sigma^2\, e^{-\beta|t-t'|} \quad . \tag{2.138}$$

The configuration process X_t is a differentiable Gaussian process of mean $\tilde{m}(t)$ and covariance $r(t,s)$

$$\tilde{m}(t) = x_0 + \frac{1-e^{-\beta t}}{\beta}\, v_0 \tag{2.139}$$

$$r(t,s) = \sigma^2 \min(t,s) + \frac{\sigma^2}{2\beta}\,(-2 + 2e^{-\beta t} + 2e^{-\beta s} - e^{-\beta|t-s|} - e^{-\beta(t+s)}) \tag{2.140}$$

then the variance is given by

$$\alpha(t) = \mathbb{E}\,((x_t - \tilde{m}_t)^2) = \frac{\sigma^2}{2\beta}\,(2\beta t - 3 + 4e^{-\beta t} - e^{-2\beta t}) \tag{2.141}$$

and the probability density takes the form

$$\tilde{p}(x_0,t_0,x,t) = \frac{1}{\sqrt{2\pi\alpha\,(t-t_0)}}\; e^{-\frac{\left(x - x_0 - \frac{v_0}{\beta}\,(1-e^{-\beta(t-t_0)})\right)^2}{2\alpha(t-t_0)}} \tag{2.142}$$

But X_t is not a Markov process, as a process in $\mathbb{R}$. Indeed the covariance does not factorize $\pi(t,s) \neq f(t)g(s)$ (see e.g. [46] and [100,a]). The invariant measure for this process is the Lebesgue measure on $\mathbb{R}$. Let us remark that if we let $\beta \to \infty$, σ^2 being fixed $\tilde{p}(x_0,t_0,x,t)$ converge to the transition probability of a Wiener process with variance σ^2. Then the Ornstein-Uhlenbeck process x_t converges in distribution to the Wiener process starting at x_0 with variance σ^2.

Let us now compute the kinematical quantities associated with the Ornstein-Uhlenbeck process with invariant measure

$$d\mu(x,v) = \frac{1}{\sqrt{\pi\beta\sigma^2}}\; e^{-\frac{v^2}{\beta\sigma^2}}\, dv\, dx \tag{2.143}$$

$$D_+x_t = v_t \;, \qquad D_+v_t = -\,\beta v_t \tag{2.144}$$

$$\frac{1}{2}(D_+ - D_-)x_t = u^x(x,v,t) = 0, \ \frac{1}{2}(D_+ - D_-)v_t = u^v(x,v,t) = -\beta v$$

$$D_-x_t = v_t \ , \quad D_-v_t = \beta v_t \tag{2.145}$$

and the stochastic acceleration $a = \frac{1}{2}(D_+D_- + D_-D_+)x_t$ vanishes.

If an external force acts on the particle, the associated Ornstein-Uhlenbeck process (2.128) is given by

$$\begin{aligned} dx_t &= v_t \, dt \\ dv_t &= -\beta v_t \, dt + f(x_t)dt + \beta\sigma dW_t \end{aligned} \tag{2.146}$$

where $f(x_t)$ is a force (the mass is let equal to one) and we assume that $f(x) = -\nabla V(x)$.

Notice that we can no longer consider the velocity process by itself. Moreover, for general f it is not easy to compute explicitly the transition probability. Nevertheless, it is possible to exhibit an explicit expression for the invariant measure which can be interpreted as a probability measure if $e^{-\frac{2V}{\beta\sigma^2}}$ is in $L^1(\mathbb{R}^d, dx)$.

Indeed, the Fokker-Planck equation (2.43) for the invariant distribution $\rho(x,v)$ assumes the form

$$\frac{\beta^2\sigma^2}{2}\Delta_v\rho(x,v) + v\cdot\nabla_x\rho(x,v) - \nabla_v\cdot[(\beta v - f)\rho\ (x,v)] = 0 \tag{2.147}$$

and admits the following solution

$$\rho(x,v) = N\, e^{-\frac{2}{\beta\sigma^2}[\frac{1}{2}v^2 + V]} \tag{2.148}$$

where N is a normalization constant which depends on V .

In this case, the kinematical quantities are given by

$$\begin{aligned} &D_+x_t = v_t \ , \quad D_+v_t = -\beta v_t + f(x_t) \\ &u^x(x,v) = 0 \ , \quad u^v(x,v) = -\beta v \\ &D_-x_t = v_t \ , \quad D_-v_t = \beta v_t + f(x_t) \end{aligned} \tag{2.149}$$

and the stochastic acceleration is just

$$a^x = \frac{1}{2}(D_+D_- + D_-D_+)x_t = f(x_t) \ . \tag{2.150}$$

Finally, let us remark that the Smoluchowski process

$$dy_t = \frac{f(y_t)}{\beta} dt + \sigma dW_t \tag{2.151}$$

is a Markovian approximation in configuration space of the Ornstein-Uhlenbeck process in the following sense [106], [90.b]:

Define $b(x,t) = \frac{f(x)}{\beta}$, let x,v be the solution of the coupled equations

$$\begin{aligned} dx_t &= v_t dt \\ dv_t &= -\beta v_t dt + \beta b(x_t)dt + \beta\sigma dW_t \end{aligned} \tag{2.152}$$

with initial conditions x_0, v_0 . Let Y be the solution of

$$dY_t = b(Y_t)dt + \sigma dW_t \ . \tag{2.153}$$

For all v_0 , with probability one holds

$$\lim_{\beta\to\infty} X_t = Y_t \tag{2.154}$$

uniformly for $t \in [0,T]$, for b and σ fixed .

III. NELSON STOCHASTIC DYNAMICS - NEWTONIAN PROCESSES

III.1 Stochastic Newton Law

Each mechanics consists of two parts: kinematics and dynamics. In the context at hand, the kinematical part of our theory, developed in the previous chapter, is to explain what we mean by "moving", and the dynamical part is to explain what we mean by "influence of an external force". The dynamics of stochastic mechanics can be given by specifying the acceleration of the diffusion, the Newton law, or through a variational principle. In this chapter, let us consider the first approach, the variational principle being studied in Chapter V.

If some dynamical law specifies the stochastic acceleration of a diffusion process, then the coupled system of non-linear equations for the osmotic velocity u and the current velocity v,

$$\frac{\partial u}{\partial t} = -\nabla\left(\frac{\sigma^2}{2}\nabla\cdot v + v\circ u\right) \tag{3.1}$$

$$\frac{\partial v}{\partial t} = a + (u\cdot\nabla)u - (v\cdot\nabla)v + \frac{\sigma^2}{2}\Delta u \tag{3.2}$$

gives us the time evolution of the infinitesimal characteristics of the diffusion and therefore, as we will discuss in chapter IV, determines it [22,a,b], [115,a,b].

Now, let ξ_t be the position of a particle of mass m at time t. The fundamental law of the non-relativistic dynamics is Newton's law $F = ma$: the force acting on a particle is the product of the mass m by the acceleration of the particle. Following Nelson's original dynamical approach [90,a,b] we introduce a stochastic analogue of the second Newton law.

In the case where the position ξ_t of the particle is described by a Markovian diffusion process, then the Nelson-Newton law (Newton law in mean) states that

$$\frac{m}{2}\,(D_+D_- + D_-D_+)\,\xi_t = F(\xi_t,t) \tag{3.3}$$

where $F(x,t)$ is an external (deterministic) force field acting on the particle.

As starting point for the stochastic mechanics, the stochastic Newton law (3.3) is not entirely satisfactory, since the direct probabilistic meaning of the stochastic acceleration is not yet well-understood. Moreover, the stochastic Newton law appears more as a constraint on the drift than as a fundamental law. We come back to this problem in the study of stochastic variational principle (Chapter V).

In the following, we call Newtonian diffusion a Markovian diffusion process for which the drift is determined by the stochastic Newton law (3.3).

III.2 Conservative Newtonian Diffusion Processes

As in classical Newtonian mechanics, an important special case is the one of conservative forces, where the force F derives from a potential V, which depends only on the position

$$F(x) = -\nabla V(x). \tag{3.4}$$

Hence the Newton law (3.3) rewrites

$$\frac{m}{2}(D_+D_- + D_-D_+)\xi_t = -\nabla V(\xi_t). \tag{3.5}$$

Moreover, if we assume that the current velocity defined by (2.57) is also a gradient*

$$v(x,t) = \nabla S(x,t) \tag{3.6}$$

the right hand side of (3.2) becomes also a gradient

$$\frac{\partial v}{\partial t} = -\nabla\left(\frac{1}{2}v^2 - \frac{1}{2}u^2 - \frac{\sigma^2}{2}\nabla\cdot u + \frac{1}{m}V\right). \tag{3.7}$$

(We have used the fact that if v is a gradient $(v\cdot\nabla)v = \frac{1}{2}(\nabla v)^2$.)

Conversely, if the velocity v is solution of

$$\frac{\partial v}{\partial t} + (v\cdot\nabla)v = \nabla\left(\frac{1}{2}u^2 + \frac{\sigma^2}{2}\nabla\cdot u - \frac{1}{m}V\right) \tag{3.8}$$

with initial condition v_o such that $v_o(x) = \nabla S_o(x)$ then the solution v of (3.8) is always a gradient.

A Markovian process such that (3.5) and (3.6) are satisfied will be called a conservative Markovian process.

Notice that the "conservative" processes defined here are qualita-

---

*Now S has the dimension of an action by unit mass.

tively different from the "dissipative" diffusion processes such those studied in Section II, §5, as Wiener process, Brownian Bridge, ...

The problem of constucting a (Markovian) diffusion process with a given initial density and given generator is well-known in the case where the drift b_+ is sufficiently smooth [22,a,b],[68],[90,c]. That is when the operator $b_+\cdot\nabla$ can be seen as a "small perturbation" of the Laplacian. In this case the measure on path space is given by a Cameron-Martin [21] Girsanov [56] formula, and is an absolutely continuous transformation of the Wiener measure with the given initial density.

In the case of conservative Newtonian diffusion process the drift b_+ is determined through the stochastic Newton law and $b_+\circ\nabla$ is no longer a "small perturbation". Before discussing the construction of a diffusion process associated with a singular drift (Chapter V), let us discuss the properties of conservative Newtonian diffusions.

III.3 Mechanics of Conservative Newtonian Process

In this section, we always assume the existence of a conservative Newtonian process ξ_t with a sufficiently smooth density of probability $\rho(x,t)$.

For such a process we can "linearize" the dynamical equations (3.1) and (3.2) by introducing the complex function ψ defined as

$$\psi(x,t) = \sqrt{\rho(x,t)}\, e^{\frac{i}{\sigma^2} S(x,t)} \tag{3.9}$$

where S is given by (3.6).

First, let us suppose that the density ρ is strictly positive. Log ρ is well-defined, consequently the osmotic velocity is finite. Setting

$$R = \frac{1}{2} \log \rho \tag{3.10}$$

ψ rewrites

$$\psi(x,t) = e^{R(x,t) + \frac{i}{\sigma^2} S(x,t)} \tag{3.11}$$

and

$$|\psi(x,t)|^2 = e^{2R(x,t)} = \rho(x,t). \tag{3.12}$$

Taking into account (3.4), (3.5) and the fact that the osmotic velocity is given by

$$u = \sigma^2 \nabla R \tag{3.13}$$

equation (3.7) rewrites

$$\frac{\partial S}{\partial t} = \frac{\sigma^4}{2} [\Delta R + (\nabla R)^2] - \frac{1}{2} (\nabla S)^2 - \frac{1}{m} V \tag{3.14}$$

and the continuity equation (3.1) takes the form

$$\frac{\partial R}{\partial t} + \frac{1}{2} \Delta S + \nabla R \cdot \nabla S = 0. \tag{3.15}$$

We recognize the Madelung fluid equation [60,a], [83] discussed in Section (I, 3,2), if we identify σ^2 with $\frac{\hbar}{m}$. Hence the function ψ verifies the "Schrödinger-like" equation

$$i \frac{\partial \psi(x,t)}{\partial t} = - \frac{\sigma^2}{2} \Delta\psi(x,t) + \frac{1}{m\sigma^2} V(x)\psi(x,t). \tag{3.16}$$

In fact, (3.14) and (3.15) are determined up to a function of time, since S and R are defined through a gradient, and (3.16) must have the general form

$$i \frac{\partial \psi(x,t)}{\partial t} = - \frac{\sigma^2}{2} \Delta\psi(x,t) + \frac{1}{m\sigma^2} V(x)\psi(x,t) + i\alpha(t)\psi(x,t) \tag{3.17}$$

where $\alpha(t)$ is real at least if (3.17) holds for all times. The reality of $\alpha(t)$ can be deduced from the fact that $\int \bar{\psi}\psi dt$ is independent of time. By an appropriate choice of S , $\alpha(t)$ can be taken zero.

Moreover, for a conservative Newtonian process the distribution of the process $\rho(x,t)$ at time t is given by

$$\rho(x,t) = |\psi(x,t)|^2 \tag{3.18}$$

where ψ is a solution of the Schrödinger equation (3.16) with initial condition

$$\psi(x,0) = \psi_0(x) \tag{3.19}$$

such that $|\psi_0(x)|^2$ is the initial distribution of the process.

Conversely, if ψ is a solution without nodes, that is $|\psi(x,t)|^2 > 0$ for any time, of the Schrödinger equation (3.16) with initial condition (3.19)

$$\psi(x,t) = e^{R + \frac{i}{\sigma^2} S} = \sqrt{\rho}\, e^{\frac{i}{\sigma^2} S} \tag{3.20}$$

with $\rho = |\psi|^2$.

Introducing now u and v by

$$u = \sigma^2 \nabla R \tag{3.21}$$

and

$$v = \nabla S \tag{3.22}$$

u and v verify equations (3.14) and (3.15). Then we can define by (2.63) the forward and backward drift b_+ and b_- and consequently a conservative diffusion process ξ_t with density $\rho(x,t) = |\psi(x,t)|^2$, the initial distribution being $|\psi(x,0)|^2$.

Moreover, the process satisfies the stochastic Newton law

$$ma(\xi_t) = -\nabla V(\xi_t). \tag{3.23}$$

To illustrate the construction of a diffusion associated with quantum evolution, we consider the simple case of the one dimensional harmonic oscillator. The evolution is described by the Schrödinger equation

$$i\hbar \frac{\partial}{\partial t} \psi = -\frac{\hbar^2}{2m} \frac{\partial^2}{\partial x^2} \psi + \frac{1}{2} m\omega^2 x^2 \psi . \tag{3.24}$$

Let us consider the solutions given by the coherent states

$$\psi_{q_0,p_0}(x,t) = \left(\frac{\pi\hbar}{m\omega}\right)^{-1/4} \exp\left[-\frac{m\omega}{2\hbar}(x-a(t))^2 + \frac{1}{\hbar} xp(t) - \frac{i}{2\hbar} p(t)q(t) - i\frac{\omega}{2} t\right] \tag{3.25}$$

associated with the classical solution $\{q(t),p(t)\}$. More precisely, $\{q(t),p(t)\}$ is the solution of the Hamilton equation (1.28) for the classical Hamiltonian

$$H = \frac{p^2}{2m} + \frac{1}{2} m\, \omega^2 q^2 . \tag{3.26}$$

Hence

$$\dot{q} = \frac{p}{m} , \qquad \dot{p} = - m\omega^2 q . \tag{3.27}$$

The classical solution associated with initial condition (q_o, p_o) at time $t = 0$ takes the form

$$\begin{aligned} q(t) &= q_o \cos\omega t + \frac{p_o}{m\omega} \sin\omega t \\ p(t) &= -m\omega q_o \sin\omega t + p_o \cos\omega t . \end{aligned} \tag{3.28}$$

The stochastic process ξ_t associated to the coherent state $\psi_{q_o,p_o}(x,t)$, has the density

$$p(x,t) = \left(\frac{\pi\hbar}{m\omega}\right)^{-1/2} \exp\left[- \frac{m\omega}{\hbar} (x-q(t))^2\right] \tag{3.29}$$

and the function S is given by the following expression

$$S(x,t) = \frac{1}{m}(xp(t) - \frac{1}{2} p(t)q(t) - \frac{1}{2} \hbar\omega t) \tag{3.30}$$

where we have chosen

$$\sigma^2 = \frac{\hbar}{m} . \tag{3.31}$$

Using eqs. (3.21) and (3.22), we deduce

$$u(x,t) = -\omega(x-q(t)) \tag{3.32}$$

and

$$v(x,t) = \frac{1}{m} p(t) \tag{3.33}$$

and by (2.63) we obtain for the drift terms

$$b_\pm(x,t) = v(x,t) \pm u(x,t) = \frac{1}{m} p(t) \pm \omega(x-q(t)). \tag{3.34}$$

Therefore, the associated stochastic equation for the process ξ_t is given by

$$d\xi_t = \left[\frac{1}{m} p(t) - \omega(\xi_t - q(t))\right] dt + \sqrt{\frac{\hbar}{m}}\, dW_t \tag{3.35}$$

where W_t is the Wiener process with variance 1 .

To solve this equation, let us introduce the process ξ_t^o such that

$$\xi_t = q(t) + \xi_t^o . \tag{3.36}$$

From (3.29) we deduce that the distribution of ξ_t^o is invariant and given by

$$\rho_o(x) = \left(\frac{\pi\hbar}{m\omega}\right)^{-1/2} e^{-\frac{m\omega}{\hbar}x^2} \tag{3.37}$$

and from (3.35) we see that the process ξ_t^o verifies the stochastic differential equation

$$d\xi_t^o = -\omega\xi_t^o \, dt + \frac{\hbar}{m} dW_t . \tag{3.38}$$

Therefore, the process ξ_t^o is just the oscillator process described in (Section II,6). This process is associated with the ground state of the harmonic oscillator

$$\psi^o(x,t) = \left(\frac{\pi\hbar}{m\omega}\right)^{-1/4} \exp\left[-\frac{m\omega}{2\hbar} x^2 + i\,\frac{\omega}{2}\,t\right] . \tag{3.39}$$

Notice that in this example the density $\rho(x,t)$ is strictly positive for any time. In the general case, the condition $\rho_o(x) = |\psi_o(x)|^2$ strictly positive at time zero does not assure the strict positivity of the density $\rho(x,t) = |\psi(x,t)|^2$ for futur time $(t > 0)$. This fact can be illustrated in the case of the harmonic oscillator by choosing as initial state

$$\psi_o(x) = \frac{1}{\sqrt{2}} (\phi_o(x) + i\phi_1(x)) \tag{3.40}$$

where $\phi_o = \pi^{-1/4} e^{-x^2/2}$ is the ground state and $\phi_1 = (\frac{\pi}{4})^{-1/4} xe^{-x^2/2}$ the first excited state. At time $t = 0$ the distribution $\rho_o(x) = |\psi_o(x)|^2$ is strictly positive, but at time t the density

$$\rho(x,t) = \frac{e^{-x^2}}{\sqrt{\pi}} \left[x^2 - \sqrt{2}\,x \sin t + \frac{1}{2}\right] \tag{3.41}$$

(where we have chosen $m = \hbar = \omega = 1$) can vanish at time $t = \pm \frac{\pi}{2}$ (mod 2π).
Therefore, a node appears in $x = \frac{1}{\sqrt{2}}$ or $x = -\frac{1}{\sqrt{2}}$ at each half period respectively!

Hence it is natural to consider the case where $\rho(x,t)$ has a non-trivial nodal set and vanishes for some values of x and t . In this case, log ρ and, consequently, the osmotic velocity $u = \frac{\sigma^2}{2}\frac{\nabla\rho}{\rho}$ are only defined on the complement of the nodes.

In Chapter IV we will show that under rather large conditions such a process with singular drift exists and that the nodes are never attained. In other words, the nodes act as an impenetrable barrier.

III.4 Conservative Newtonian Processes with Stationary Distribution

In the previous example, we have considered the oscillator process with stationary distribution. Let us now study the general feature of processes with stationary distribution i.e.

$$\frac{\partial\rho}{\partial t}(x,t) = 0 . \tag{3.42}$$

Hence ρ does not depend on time

$$\rho(x,t) = \rho_o(x) . \tag{3.43}$$

As a consequence, the osmotic velocity in the region where it exists is time independent

$$u(x) = \frac{\sigma^2}{2}\frac{\nabla\rho(x)}{\rho(x)} . \tag{3.44}$$

Using the Ansatz (3.9)

$$\psi(x,t) = \sqrt{\rho(x)}\, e^{\frac{i}{\sigma^2} S(x,t)} \tag{3.45}$$

and (3.42) the continuity equation (3.1) takes the following form

$$\nabla \circ (\rho \nabla S) = 0 \tag{3.46}$$

and the equation (3.7) rewrites

$$\frac{\partial S}{\partial t} - \frac{\sigma^4}{2}\Delta\log\rho + \frac{1}{2}(\nabla S)^2 - \frac{\sigma^4}{8}(\nabla\log\rho)^2 + \frac{V}{m} = 0. \tag{3.47}$$

The system of coupled partial differential equations (3.46), (3.47) with initial condition

$$S(x,0) = S_o(x) \tag{3.48}$$

admits a solution of the following form

$$S(x,t) = -\frac{E}{m} t + S_o(x) \tag{3.49}$$

if and only if the equation

$$\left(\frac{\sigma^2}{2} \Delta - V\right) \phi = E \phi \tag{3.50}$$

admits solutions of the form

$$\phi(x) = \sqrt{\rho(x)}\, e^{\frac{i}{\sigma^2} S_o(x)} . \tag{3.51}$$

Indeed, for $S(x,t)$ given by (3.49), (3.50) rewrites

$$\frac{E}{m} = -\frac{\sigma^4}{4} \Delta \log\rho - \frac{\sigma^4}{8} (\nabla \log\rho)^2 + \frac{1}{2} (\nabla S_o)^2 - \frac{1}{m} V \tag{3.52}$$

for x such that $\rho \neq 0$.

The continuity equation becomes

$$\Delta S_o + \frac{\nabla\rho}{\rho} \nabla S_o = 0. \tag{3.53}$$

Hence

$$\phi(x) = \sqrt{\rho(x)}\, e^{\frac{i}{\sigma^2} S_o(x)}$$

is solution of (3.50) and

$$\psi(x,t) = \phi(x)\, e^{-i \frac{E}{m\sigma^2} t} \tag{3.54}$$

is a stationary solution of (3.16).

Conversely, given (3.54), then (3.51) is solution of (3.50). By splitting in real and imaginary parts, we get the eqs. (3.47) and (3.48).

Moreover, the current velocity does not depend on time

$$v(x) = \nabla S_o(x) \tag{3.55}$$

and the same occurs for the forward and backward drifts $b_+(x)$ and $b_-(x)$.

Now, the associated process ξ_t verifies the stochastic differential equation

$$d\xi_t = b_+(\xi_t)dt + \sigma dW_t \ . \tag{3.56}$$

It is a homogeneous Markovian process with stationary distribution, hence ξ_t is a stationary process.

Let us remark that only the drift associated with the strictly positive ground state will be non-singular. Each excited state since orthogonal to the ground state, will have a non-trivial nodal set. An interesting situation is the case where the nodal surfaces split the configuration space in closed disjoint domains. The trajectories of the process are trapped in one of the domain. Such a diffusion process furnishes a model of confinement by impenetrable barrier. We can understand this property in a heuristic way. The osmotic velocity $u(x)$ satisfies (2.62), $u(x) = \frac{\sigma^2}{2} \nabla \log\rho(x)$ hence has a singularity on the surface $\rho(x) = 0$. However, the gradient of a function points towards the region where the function increases. Then $u(x)$, hence $b_+(x)$ points outside of the nodal surface. If we remember that the heuristic interpretation of $b_+(x)$ is the mean velocity of particles which leave the point x, then typical trajectories of ξ_t are repelled by the nodal surface. In the next section, we will consider the stationary case and prove that the nodal set is indeed never reached.

III.5 Unattainability of the Nodes for Stationary Diffusion Processes

In this section we limit our consideration to the case of a stationary Markov diffusion process ξ_t solution of the stochastic differential equation

$$d\xi_t = b_+ (\xi_t) \ dt + dW_t \tag{3.57}$$

where W_t is a Wiener process in $\mathbb{R}^1$ with convariance matrix $\mathbb{1} t$ and with a stationary density $\rho \geq 0$, $\rho \in C^\infty$, $\rho(x) > 0$ a.e. (Notice that the L^1_{loc} character of ρ does not exclude densities which are not probability densities but define stationary measures).

Moreover we assume in the following that the drift b_+ is a gradient, namely

$$b_+(x) = \frac{1}{2} \nabla \log\rho \ . \tag{3.58}$$

The stationarity implies that the current velocity vanishes, $v = 0$ and $b_+(x) = -b_-(x) = u(x)$ then ξ_t is a symmetric diffusion process (see Appendix).
The diffusion process we consider is associated with the generator

$$A = \frac{1}{2}\,\Delta + \frac{1}{2}\,(\nabla\cdot\log\rho)\cdot\nabla \tag{3.59}$$

outside the open set N_ε defined by

$$N_\varepsilon = \{x \in \mathbb{R}^d \,|\, \rho < \varepsilon\ ,\ \varepsilon > 0\}\ . \tag{3.60}$$

The drift b_+ and the generator A are well defined and we consider the process in the complement of N_ε. Our aim is to prove that, for a large class of distributions ρ, with probability one

$$\lim_{\varepsilon\downarrow 0} \tau_{N_\varepsilon} = \infty \tag{3.61}$$

where

$$\tau_{N_\varepsilon} = \inf\,\{t > 0 \,|\, \xi_t \in N_\varepsilon\} \tag{3.62}$$

is the first hitting time of N_ε (see Appendix).

Instead of considering the linear operator A defined in (3.59) let us introduce the associated bilinear form E

$$E(f,g) = -(A\,f,\ g)_{L^2(\mathbb{R}^d,\rho\,dx)}\ ; f,g \in C_0^\infty\,(\mathbb{R}^d) \tag{3.63}$$

the so called (symmetric) energy form

$$E(f,g) = \frac{1}{2}\int_{\mathbb{R}^d} \nabla f\cdot\nabla g\ \rho\ dx\ ;\ f,g \in C_0^\infty\,(\mathbb{R}^d)\ . \tag{3.64}$$

For $\rho \equiv 1$ we recover just the classical Dirichlet integral.

In contrast to (3.53), formula (3.64) allows discontinuities, singularities and degeneracy of ρ.

Let us now assume that the following condition on ρ is verified

(#) $\rho = |\psi|^2$ with $|\psi| \in H^1_{loc}\,(\mathbb{R}^d)$ the first Sobolev space (i.e. $|\psi|$ and $\nabla|\psi|$ belong to $L^2_{loc}\,(\mathbb{R}^d)$) and $|\psi| > 0$ a.e.

This condition (#) is equivalent to $\rho \in L^1_{loc}\,(\mathbb{R}^d)$, $\frac{\nabla\rho}{\rho} \in L^2_{loc}(\mathbb{R}^d,\rho dx)$ and $\rho > 0$ a.e.

Under the condition (#) the symmetric form (3.64) defined on $C_0^\infty(\mathbb{R}^d)$ is closable in $L^2(\mathbb{R}^d, \rho\, dx)$ in the following sense :

The symmetric form E is called closable in $L^2(\mathbb{R}^d, \rho\, dx)$ if $E(f_n - f_m;\, g_n - g_m) \to 0$, $n, m \to \infty$, $|(f_n, f_n)_{L^2(\mathbb{R}^d, \rho\, dx)}| \to 0$, $n \to \infty$ implies $E(f_n, f_n) \to 0$, $n \to \infty$. To be closable is equivalent with the following property: $L^2(\mathbb{R}^d, \rho\, dx)$ is complete with respect to the metric E_1

$$E_1(f,g) = E(f,g) + (f,g)_{L^2(\mathbb{R}^d, \rho\, dx)} \tag{3.65}$$

Indeed condition (#) implies that if $\{f_n\}_{n \in \mathbb{N}}$, $f_n \in C_0^\infty(\mathbb{R}^d)$ is a sequence such that $(f_n, f_n)_{L^2(\mathbb{R}^d, \rho\, dx)} \to 0$, $n \to \infty$ then $E(f_n, g) \to 0$, $n \to \infty$ for any $g \in C_0^\infty(\mathbb{R}^d)$ and the closability follows.

Moreover under the condition (#) there exists a unique diffusion process ξ_t on $\mathbb{R}^d$ associated to the form E up to a set of zero capacity [8],[50,a,b],[94]. More precisely there is a unique positive self adjoint operator H associated to the closure $\bar{E}$ of E and such that the domain of H $D(H) \subset D(\bar{E})$ (the domain of $\bar{E}$) and $\bar{E}(f,g) = (H f, g)_{L^2(\mathbb{R}^d, \rho\, dx)}$, $\forall f \in D(H)$, $g \in D(\bar{E})$.
Given the positive self adjoint operator H we can construct a diffusion process with values in $\mathbb{R}^d$, with transition function P_t symmetric in $L^2(\mathbb{R}^d, \rho\, dx)$ such that

$$P_t f = e^{-tH} f \;, \quad f \in L^2(\mathbb{R}^d, \rho\, dx) \;. \tag{3.66}$$

Proof of these facts can be found in [8], [50,a], [100,c].

The capacity $\mathrm{Cap}(\mathcal{O})$ is defined for an open set $\mathcal{O}$ (w.r.t. the measure ρdx) by

$$\mathrm{Cap}(\mathcal{O}) = \inf\{E_1(f,f) : f \in L_{\mathcal{O}} = \{f \in D(E)\ f \geq 1, \text{a.e.on } \mathcal{O}\}\} \tag{3.67}$$

where

$$E_1(f,f) = E(f,f) + (f,f)_{L^2(\mathbb{R}^d, \rho\, dx)} \tag{3.68}$$

and $D(E)$ is the domain of the (regular) Dirichlet form E (the smallest closed extension of E, denoted by E again). (The infimum is taken to be $+\infty$ if $L_{\mathcal{O}} = \emptyset$.) For an arbitrary set B the capacity is defined as

$$\mathrm{Cap}(B) = \inf\{\mathrm{Cap}(\mathcal{O}) : \mathcal{O} \text{ open } \mathcal{O} \supset B\}. \tag{3.69}$$

The minimum in (3.67) is assumed by a unique function $e_{\mathcal{O}} \in L_{\mathcal{O}}$ which minimizes E_1 , this function being the "equilibrium potential". One has $0 \leq e \leq d$ and $\mathrm{Cap}(\mathcal{O}) = E[e_{\mathcal{O}}, e_{\mathcal{O}}] + (e_{\mathcal{O}}, e_{\mathcal{O}})$.
The function $e_{\mathcal{O}}$ is equal to the hitting probability

$$e_{\mathcal{O}}(x) = \mathbb{E}_x(e^{-\tau_{\mathcal{O}}}) \tag{3.70}$$

where $\tau_{\mathcal{O}}$ is the first time the process starting from x at time 0 hits the set $\mathcal{O}$ (see [50,a,b]. From this it follows that an open set $B \subset \mathbb{R}^d$ has zero capacity iff the probability that $P(\xi_t \in B$ for some $t \geq 0 \mid \xi_0 = x) = 0$ for any $x \in \mathbb{R}^d \setminus B$.

Remark For the classical Dirichlet integral $(\rho = 1)$ the equilibrium potential of $\mathcal{O}$ is given by a function which is harmonic on $\mathbb{R}^d \setminus \mathcal{O}$ and identically 1 on $\mathcal{O}$.

Consider $|\psi| \in H^1_{loc}(\mathbb{R}^d)$, $|\psi| > 0$ a.e. and such that $|\psi|$ is locally a E-quasi continuous function (namely for any $\varepsilon > 0$ there exists an open set B with $\mathrm{Cap}(B) < \varepsilon$ such that the restriction of $|\psi|$ to $\mathbb{R}^d \setminus B$ is continuous) with finite Dirichlet integral (that is just the Dirichlet form for $\rho = 1$).

Let

$$N = \{x \in \mathbb{R}^d \mid \psi(x) = 0\} \tag{3.71}$$

and ξ_t be the diffusion process associated to E with $\rho = |\psi|^2$.

Theorem 3.1: Under the above condition, if $|\psi|$ is locally bounded from above, then

$$P[\tau_N < +\infty] = 0 \qquad \text{q.e.,} \quad x \in \mathbb{R}^d \tag{3.72}$$

where τ_N is the first hitting time of N

$$\tau_N = \inf\{t > 0 \mid \xi_t \in N\} \tag{3.73}$$

and q.e. (quasi everywhere means "excepted on a set of capacity 0").

Proof: Using the relation between the capacity and hitting time (3.70) it suffices to show (see [50,a,b]) that

$$\mathrm{Cap}(N \cap B_r) = 0 \ , \ B_r = \{x \in \mathbb{R}^d \mid |x| < r\} \ \forall r > 0.$$

Consider a function $f \in C_0^\infty(\mathbb{R}^d)$ such that $0 \leq f \leq 1$ with $f = 1$ on B_r and $f = 0$ on B_{r+1}^c (the complement of B_{r+1}) and set

$$g(x) = \log|\psi(x)| \cdot f(x)$$

$$g_\varepsilon(x) = \log[|\psi(x)| \vee \varepsilon] \cdot f(x) \qquad \varepsilon > 0 ,$$

with $f \vee g = \max(f,g)$.
The assumption (#) on $|\psi|$ implies that g_ε is E-quasi continuous. Since $|\psi| \in H^1_{loc}(\mathbb{R}^d)$, $|\psi|$ belongs to $L^p_{loc}(\mathbb{R}^d)$ for $p > 2$ and $\log|\psi| \in L^2(\mathbb{R}^d, |\psi|^2 dx)$. It follows from this that $g \in L^2(\mathbb{R}^d, |\psi|^2 dx)$ and

$$\int_{\mathbb{R}^d} |\nabla g|^2 |\psi|^2 dx \leq 2 \int_{\mathbb{R}^d} |\nabla|\psi||^2 f^2 dx + 2 \int_{\mathbb{R}^d} |\nabla f|^2 (\log|\psi|)^2 |\psi^2| dx < +\infty.$$

The same property holds for g_ε. Moreover, we have

$$\lim_{\varepsilon \downarrow 0} \int_{\mathbb{R}^d} |\nabla g_\varepsilon|^2 \; |\psi|^2 dx = \int_{\mathbb{R}^d} |\nabla g|^2 |\psi|^2 dx .$$

By regularization it is easy to see that g_ε is a E_1^ρ-limit of C_0^∞-function and hence g_ε is a quasi-continuous function on the Dirichlet space $D(E^\rho)$. We can therefore write (by using a Chebyshev's type inequality)

$$\mathrm{Cap}(|g_\varepsilon| > \lambda) \leq \frac{1}{\lambda^2} E_1(g_\varepsilon, g_\varepsilon) \tag{3.74}$$

and letting $\varepsilon \downarrow 0$ we obtain

$$\mathrm{Cap}(|\log\psi|(x)|| \; |f(x)| > \lambda) \leq \frac{1}{\lambda^2} \left[\int |\nabla g|^2 |\psi^2| dx + \int g^2 |\psi|^2 dx \right] \underset{\lambda \to \infty}{\to} 0 \tag{3.75}$$

and since $\mathrm{Cap}(N \cap B_r)$ is smaller than the left hand side of the above equation, the theorem is proved.

Remark: If the condition (#) is not verified, the paths of the process X_t can reach or cross the nodal surface of $|\psi|$. (See e.g. [50,b].)

Consider the case where $d = 1$, $\rho \in L^1_{loc}(\mathbb{R})$ non negative such that $\inf_{a<x<b} \rho(x) > 0$ if $0 < a < b < +\infty$ then $\rho(0)$ can take the value zero. The origin is unattainable from the right if and only if

$$\int_0^b \frac{dx}{\rho(x)} = +\infty \ , \ b > 0 \tag{3.76}$$

(see eg [8]).

If ρ satisfies the condition (#) then we have

$$\int_0^b |\psi|'^2 \, dx < +\infty \tag{3.77}$$

and using Schwartz' inequality we conclude that.

$$\int_\varepsilon^b |\psi|'^2 \, dx \int_\varepsilon^b \frac{1}{|\psi|^2} \, dx \geq \left[\log|\psi|(b) - \log|\psi|(\varepsilon) \right]^2 \tag{3.78}$$

from which it follows that the right hand side goes to $+\infty$ as $\varepsilon \downarrow 0$, which shows that indeed the origin is unattainable.

III.6 Diffusion in an External Electromagnetic Field

In the previous sections, we have considered the case of a diffusion in presence of an external conservative force. Let us now consider the case where an external electromagnetic field is present. The dynamical law for a particle of mass m takes the form

$$ma(x,t) = F_1(x,t) + F_2(x) \tag{3.79}$$

where a is the acceleration defined in (2.94). We assume the force F_1 to be the Lorentz force

$$F_1(x,t) = q[E(x,t) + v(x,t) \times B(x,t)] \tag{3.80}$$

where q is the charge of the particle. E and B are, respectively, the electric and magnetic field. The velocity $v(x,t)$ is chosen to be the current velocity (2.60). This last assumption assures the usual behaviour under time reversal. Moreover, $F_2(x) = -\nabla U(x)$ where U is a scalar potential which accounts for the other possible forces.

Let A be the vector potential and ϕ the scalar potential, then

$$\begin{aligned} B &= \nabla \times A \\ E &= -\frac{\partial A}{\partial t} - \nabla\phi \ . \end{aligned} \tag{3.81}$$

If furthermore we make the gauge invariant assumption that there exists a function $S(x,t)$ such that

$$mv + qA = \nabla S \tag{3.82}$$

we can rewrite the coupled system of non-linear equations (3.1),(3.2)

$$\frac{\partial S}{\partial t} - \frac{m\sigma^4}{2}(\nabla \log \rho^{1/2})^2 - \frac{m\sigma^4}{2}\Delta \log \rho^{1/2} + \frac{1}{2m}(\nabla S - qA)^2 + U + q\phi = 0 \tag{3.83}$$

$$\frac{\partial \rho^{1/2}}{\partial t} + \nabla \rho^{1/2} \cdot \nabla S + \rho^{1/2}\Delta S - qA\cdot\nabla\rho^{1/2} = 0 \tag{3.84}$$

the latter equation is the continuity equation.

If following Section (III.3) we make the Ansatz

$$\psi(x,t) = \rho^{1/2}(x,t)\, e^{\frac{i}{m\sigma^2} S(x,t)} \tag{3.85}$$

ψ solves the linear equation

$$im\sigma^2 \frac{\partial \psi}{\partial t}(x,t) = \frac{1}{2m}(-im\sigma^2\nabla - qA)^2\psi(x,t) + (U(x,t) + q\phi(x,t))\psi(x,t) \ . \tag{3.86}$$

Vice versa, given a solution of this Schrödinger type equation, then the formula (3.75) defines the density ρ and the drift b_+ of a diffusion process.

III.7 Newtonian Diffusion on Riemannian Manifold

Let $(\mathbb{M},g)$ be a smooth oriented d-dimensional Riemannian manifold. In a local coordinate system in $\mathbb{M}$, the components g_{ij} of the metric tensor g and its contravariant components g^{ij} verify

$$g^{ij} g_{jk} = \delta^i_k \tag{3.87}$$

where the sum is taken on repeated index.
Let X_t, $t \in T \subset \mathbb{R}_+$, be a diffusion process with values in $\mathbb{M}$. The analytic description of X_t is given by its infinitesimal generator A_t, which assumes, on $C_o^\infty(\mathbb{M})$ functions the following form

$$A_t = \frac{1}{2}\Delta_g + b_+ \cdot \text{grad} \tag{3.88}$$

where $b_+ \cdot \text{grad} = b_+^i \frac{\partial}{\partial x^i} = g_{ij}\, b_+^i \frac{\partial}{\partial x_j}$ in local coordinates, b_+ is a non random C^∞ -vector field, the drift, which might depend explicitly on the time t.

Δ_g is the Laplace-Beltrami operator on $\mathbb{M}$. In local coordinates we have

$$\Delta_g f = \sqrt{g}\, \partial_i (\sqrt{g}\, g^{ij}\, \partial_j f) \tag{3.89}$$

for any scalar function f, $\partial_i = \frac{\partial}{\partial x^i}$ and

$$g = \det[g_{ij}] . \tag{3.90}$$

The connection between b_+ and the process X_t is that $b_+(X_t,t)$ is the (mean) forward derivatives of X_t at time t in the sense that

$$b_+^i (x,t) = \lim_{\Delta t \downarrow 0} (\Delta t)^{-1}\, \mathbb{E}\,[Y^i{}_{\Delta t} \mid X_t = x] \tag{3.91}$$

where $\mathbb{E}\,[\cdot \mid X_t = x]$ means conditional expectation with respect to $X_t = x$. $Y^i_{\Delta t}$ is the vector attached to X_t tangent to geodesics from X_t to $X_{t+\Delta t}$, with length $|Y_{\Delta t}|$ equal to the geodesics distance of $X_{t+\Delta t}$ and X_t. $b_+(X_t,t)$ is also the forward stochastic derivative in the sense of [29],[40],[85,f],[90,e].

Let $dx = \sqrt{g}\, dx^1 \ldots dx^d$ be the Riemannian volume element on $\mathbb{M}$. Due to the assumption we know that there exists a smooth density $\rho(x,t)$ of the law of X_t with respect to dx, i.e. $dP(X_t \in dx) = \rho(x,t)dx$.

Let $f \in C_0^\infty(M)$, then $\mathbb{E}\,[f(X_t)] = \int_M f(x)\rho(x,t)dx$ and $\frac{d}{dt}\, \mathbb{E}\,[f(X_t)] = \int_M f(x) \frac{\partial\rho}{\partial t}(x,t)dx$. On the other hand, by the definition of L_t, the left hand side is equal to $\int_M (A_t f)\ (x,t)\rho(x,t)dx$. By partial integrations we arrive at the Kolmogorov forward equation (Fokker-Plack equation)

$$\frac{\partial}{\partial t}\, \rho = \frac{1}{2}\, \Delta_g \rho - \operatorname{div}(b_+\rho) , \tag{3.92}$$

where $\operatorname{div} V = \sqrt{g}\, \partial_i \sqrt{g}\, V^i$ for any smooth vector field V. $A_t^* f$ defined on C^2 function by $A_t^* f = \frac{1}{2}\, \Delta_g f - \operatorname{div}(b_+ f)$ being the adjoint of A_t.

Let us now denote by $\check{X}_t$, $t \in -T$, the time reversed process to X_t, i.e. $\check{X}_{-t}$ has the same law as X_t. It is well-known, see e.g.

[41] and section II.5 that $\check{X}_t$ is again a Markov process with infinitesimal generator

$$\check{A}_t \equiv \frac{1}{2} \Delta_g - b_- . \text{ grad} \tag{3.93}$$

b_-^i being the "backward drift" defined, for $t \in T$, by

$$b_-^i(x,t) = \lim_{\Delta t \downarrow 0} (\Delta t)^{-1} E [Y_{-\Delta t}^i \mid X_t = x] \tag{3.94}$$

with $Y_{-\Delta t}^i$ defined as $Y_{\Delta t}^i$ with $-\Delta t$ replacing Δt . Then b_- is the backward stochastic derivative of X_t . By the same procedure as above, one arrives at the Fokker-Planck equation for the reversed process $t \in T$:

$$- \frac{\partial}{\partial t} \rho = \frac{1}{2} \Delta_g \rho + \text{div}(b_- \rho). \tag{3.95}$$

As in the Euclidean case, we can introduce the osmotic velocity u and the current velocity v

$$u = \frac{1}{2} (b_+ - b_-) \tag{3.96}$$

$$v = \frac{1}{2} (b_+ + b_-) . \tag{3.97}$$

As in Chapter II, we can derive the "continuity equation"

$$\frac{\partial \rho}{\partial t} = -\text{div}(\rho v) \tag{3.98}$$

and the "osmotic equation"

$$\frac{1}{2} \Delta_g \rho = \text{div}(\rho u) . \tag{3.99}$$

Moreover, the osmotic velocity u is given by

$$u = \frac{1}{2} \text{grad} \log\rho . \tag{3.100}$$

This follows from the generalization of formula (1.52) to Riemmanian manifold. Let $f,h \in C_0^\infty (T\times M)$ then

$$\begin{aligned} 0 &= \int_T \frac{d}{dt} \mathbb{E} [f(X_t,t)h(X_t,t)]dt \\ &= \int_T \mathbb{E} \Big[(D_+ f(X_t,t))h(X_t,t)\Big]dt + \int_T \mathbb{E}\Big[f(X_t,t)D_- h(X_t,t)\Big]dt \end{aligned} \tag{3.101}$$

where $D_+ \equiv \frac{\partial}{\partial t} + \frac{1}{2}\Delta_g + b_+.\text{grad}$ (the operator of mean derivative on functions) and $D_- \equiv \frac{\partial}{\partial t} - \frac{1}{2}\Delta_g + b_-.\text{grad}$. Using partial integrations to bring D_+ to act on $h\rho$ and using Fokker-Planck's equation (3.92), we arrive easily from this to the conclusion that $-D_- = -\frac{\partial}{\partial t} - b_+.\text{grad} + \text{grad}\,(\text{Log}\rho).\text{grad} + \frac{1}{2}\Delta_g$. Using $b_- = D_-X_t$ we then get (3.100). From this equation (3.100) we have, taking the time derivative and using the continuity equation (3.98)

$$\frac{\partial u}{\partial t} = -\,\text{grad div}\, v - \text{grad}\,(v.u). \tag{3.102}$$

We shall now define the mean acceleration associated with the process X_t . To do this, we would like to have the concept of mean forward and backward derivatives of vectors on $\mathbb{M}$. The appropriate definition has been given by Dohrn and Guerra [39], [90,c]. Let $F = F^i(x,t)$ be a vector field on $\mathbb{M}$, then the mean forward derivative of F is defined by

$$D_+F(x,t) \equiv \lim_{\Delta t \downarrow 0} (\Delta t)^{-1}\, E[\tau_{X_t, X_{t+\Delta t}} F(X_{t+\Delta t}, t+\Delta t) - F(X_t,t)\,|\,X_t = x] \tag{3.103}$$

where $\tau_{y,y+\Delta y}F$ is the vector in the tangent space $T_{y+\Delta y}\mathbb{M}$ at $y+\Delta y$ obtained from the vector $F \in T_y\mathbb{M}$ by Dohrn-Guerra's stochastic parallel transport along the geodesics from y to $y+\Delta y$. We recall briefly the definition of this transport (for more details see [39]).

Let $\gamma(s)$, $s_0 \leq s \leq s_1$, be a segment of geodesics on $\mathbb{M}$. Let F^i be a vector in $T_{\gamma(s_0)}\mathbb{M}$. Let $h(t)$, $t \in [0,1]$, be a curve on $\mathbb{M}$ such that $h(0) = \gamma(s_0)$ and $\dot{h}(0) = F$. Let us transport in a Levi-Civita parallel way $\dot{\gamma}(s_0) \equiv G(t)$ along $h(t)$, getting a vector field $G(t)$. Let $\gamma_t(s)$ be the geodesic such that $\gamma_t(s_0) = h(t)$, $\dot{\gamma}_t(s_0) = G(t)$, $s < s_1$. The family of geodesics $\{\gamma_t(s),\ s_0 \leq s \leq s_1,\ t \in [0,1]\}$ is a parallel for $s = s_0$. Let $B(s) = \frac{d}{dt}\gamma_t(s)|_{t=0}$; this is a vector field along $\gamma(s)$, with $B(s_0) = F$. By definition, $\tau_{\gamma(s_0),\gamma(s)}F \equiv B(s)$ and $\tau_{\gamma(s_0),\gamma(s)}F$ differs from the Levi-Civita displacement of F by second order terms in s . This then gives, for $y = \gamma(s_0)$, $y+\Delta y = \gamma(s)$ the transport $\tau_{y,y+\Delta y}F$ needed in (2.11). One computes easily

$$D_+ = \frac{\partial}{\partial t} + b_+.\text{grad} + \frac{1}{2}\Delta_{DR} \tag{3.104}$$

$$D_- = \frac{\partial}{\partial t} + b_-.\text{grad} - \frac{1}{2}\Delta_{DR} \tag{3.105}$$

$\Delta_{DR} \equiv \Delta + R$ being the Laplace-de Rham-Kodaira Laplacian on $\mathbb{M}$, R being the Ricci tensor, acting on vectors.

Δ is given by

$$\Delta = D_i \, D^i \tag{3.106}$$

where D_i is the covariant derivative, defined on vector field F by

$$D_i F^j = \frac{\partial}{\partial x^i} F^j + \Gamma^i_{jk} \, F^k \tag{3.107}$$

and Γ^i_{jk} are the Christoffel symbols associated with the metric tensor

$$\Gamma^k_{ij} = g^{kh} \, \Gamma_{h,ij} \tag{3.108}$$

$$\Gamma_{k,ij} = \frac{1}{2} \left(\frac{\partial}{\partial x^i} g_{kj} + \frac{\partial}{\partial x^j} g_{ik} - \frac{\partial}{\partial x^k} g_{ij} \right) \tag{3.109}$$

R is the Ricci tensor, defined by

$$R_{ij} = \frac{\partial}{\partial x^j} \Gamma^k_{ki} - \frac{\partial}{\partial x^k} \Gamma^k_{ji} + \Gamma^k_j - \Gamma^l_{ki} - \Gamma^k_{kl} \, \Gamma^l_{ji} \, . \tag{3.110}$$

As in the case where $\mathbb{M} = \mathbb{R}^d$, we define the <u>mean acceleration</u> $a(X_t,t)$ by

$$a(X_t,t) \equiv \frac{1}{2}(D_+D_- + D_-D_+)X_t \, .$$

Using $D_+X_t = b_+(X_t,t)$ and $D_-X_t = b_-(X_t,t)$ we get

$$a(X_t,t) = \frac{1}{2} \, D_+b_-(X_t,t) + \frac{1}{2} \, D_-b_+(X_t,t)$$

$$= \frac{1}{2}(\partial_t + b_+ \cdot \mathrm{grad} + \tfrac{1}{2}\Delta_{DR})\frac{v-u}{2} + \frac{1}{2}(\partial_t + b_- \cdot \mathrm{grad} - \tfrac{1}{2}\Delta_{DR})\frac{v+u}{2}$$

$$= (\partial_t v + v \cdot \mathrm{grad}\, v - u \cdot \mathrm{grad}\, u - \tfrac{1}{2}\Delta_{DR} u)(X_t,t) \tag{3.111}$$

hence

$$\frac{\partial v}{\partial t} = a + u \cdot \mathrm{grad}\, u - v \cdot \mathrm{grad}\, v + \frac{1}{2}\Delta_{DR} u \, . \tag{3.112}$$

Let us also remark that a purely probabilistic description of the process is given by the solution of the stochastic differential equation (in Itô's sense)

$$dX_t = b_+(X_t,t)dt + dB_t \ , \tag{3.113}$$

which B_t the standard Brownian motion on $\mathbb{M}$, which is related to the standard Brownian motion W_t in $\mathbb{R}^d$ by the following Itô stochastic differential equation

$$dB_t^i = m^i(B_t)dt + \sigma_k^i(B_t)dW_t^k \tag{3.114}$$

m^i and σ_k^i being given in terms of the tensor metric by the equation

$$\begin{aligned} m^i &= -\frac{1}{2} g^{ik} \Gamma^i_{jk} \\ \sigma^{ik} \sigma_{kj} &= \sigma^i_j \ . \end{aligned} \tag{3.115}$$

Of course, this is not an intrinsic description, for such one see [68], [85,b].

Given b_+, we can under suitable assumptions construct X_t and get ρ, hence b_- and hence u and v satisfying (3.95), (3.99). Moreover, given X_t and its distribution ρ we can get b_+ and b_- as mean forward resp. backward derivatives and then get u,v satisfying (3.95), (3.99), a being the mean acceleration of X_t.

Now, in the same spirit as in section (3.2), we can define a conservative Newtonian diffusion. If X_t satisfies Newton's law in the mean, in the sense that there exists a positive constant m and a real-valued function V on $\mathbb{M} \times \mathbb{T}$ such that

$$\frac{m}{2} (D_+D_- + D_-D_+)X_t = - \operatorname{grad} V(X_t) \tag{3.116}$$

and such that in addition the corresponding current velocity is a gradient field

$$v(x,t) = \operatorname{grad} S(x,t) \tag{3.117}$$

we will say, as in the Euclidean case where $\mathbb{M} = \mathbb{R}^d$, that X_t is a Newtonian diffusion.

Introducing the complex function on the manifold $\mathbb{M}$

$$\psi(x,t) = \sqrt{\rho(x,t)} \quad e^{iS(x,t)} \tag{3.118}$$

it can easily be shown that ψ verifies the partial differential equation

$$\frac{\partial\psi}{\partial t}(x,t) = -\frac{1}{2}\Delta_g\psi(x,t) + V\psi(x,t) \tag{3.119}$$

where Δ_g is the Laplace-Beltrami operator on $\mathbb{M}$.
Vice versa, if ψ is solution of (3.106) and if we write $\psi = \sqrt{\rho}\, e^{iS}$ and define u and v by $u = \operatorname{grad}\log\rho$ and $v = \operatorname{grad} S$, then u and v satisfy (3.95) and (3.99). From u and v we can get in particular b_+ and thus the stochastic equation for a process X_t , the distribution of which is then for all times $\rho(x,t) = |\psi(x,t)|^2$, if at time $t = 0$ it has distribution $|\psi(x,0)|^2$. Moreover, the process satisfies Newton's equation in the mean.

Finally, let us remark that in the case of Riemannian manifold the problems of nodes can be investigated in the stationary case along the same methods as in Section (3.5). In particular, the nodes of the density ρ can never be reached by Newtonian diffusion. In the next chapter, we will give in the non-stationary case a proof of the unattainability of the nodes which works also in the case where the state space of the process is a Riemannian manifold.

IV. GLOBAL EXISTENCE FOR DIFFUSIONS WITH SINGULAR DRIFTS

IV.1 Introduction

Within the context of diffusion processes, the main bulk of the mathematical literature (see e.g. [55], [68], [104]) discusses the question of existence and uniqueness of solutions of stochastic differential equations in Itô's sense under assumptions very reminiscent of those for deterministic differential equations. Usually the coefficients of the stochastic differential equation (the so-called infinitesimal characteristics b_+ and b_-) are required to satisfy some regularity condition (such as a Lipschitz condition) to ensure local existence and uniqueness of a continuous solution and a growth condition on the drift b_+ is imposed to avoid explosions, i.e. to avoid the process of moving off to infinity within finite time. Both from a mathematical point of view and a look towards applications in other fields, such as physics or biology, it is of interest to relax the standard conditions. There are physical situations, e.g. occurring in stochastic mechanics, where one desires to construct diffusions with extremely singular drifts but well-behaved Brownian path.

In Chapter III, we have investigated the possibility of singular drift in the case of stationary processes, using the properties of Dirichlet forms. In this Section, we consider the case where ρ is not necessarily stationary and a pure probabilistic approach in terms of stopping time is used. The probabilistic approach has the advantage of exhibiting explicitly that the diffusions avoid nodes of the density.

Let us consider a wave function ψ in $\mathbb{R}^d$ satisfying the Schrödinger equation

$$\hbar \frac{\partial\psi}{\partial t} = i \frac{\hbar^2}{2m} \Delta\psi - iV\psi \tag{4.1}$$

with initial condition $\psi(x,0) = \psi_0(x)$.

We write formally

$$\psi(x,t) = e^{R(x,t)+iS(x,t)} \tag{4.2}$$

with R and S real. In this Section, we choose S without dimension, then the current velocity takes the form $v = \frac{\hbar}{m} \nabla S$.

Then we define two vector fields b_+ and b_- by

$$b_+(x,t) = \frac{\hbar}{m} [\mathrm{Re} \frac{\nabla\psi}{\psi} + \mathrm{Im} \frac{\nabla\psi}{\psi}] = \frac{\hbar}{m} (\nabla R + \nabla S) \tag{4.3}$$

$$b_-(x,t) = \frac{\hbar}{m} [-\mathrm{Re} \frac{\nabla\psi}{\psi} + \mathrm{Im} \frac{\nabla\psi}{\psi}] = \frac{\hbar}{m} (-\nabla R + \nabla S) \qquad (4.4)$$

and a non-negative function $\rho(x,t) = |\psi(x,t)|^2$. b_+ and b_- are only well-defined on the complement of the nodal set $N = \{(x,t) \in \mathbb{R}^d \times \mathbb{R} \;|\psi(x,t) = 0\}$. Using the complex conjugate equation of (4.1) it is possible, at least formally, to associate a Markov diffusion process to ρ. Indeed, an easy calculation shows that ρ must satisfy a Fokker-Planck equation, namely

$$\frac{\partial\rho}{\partial t} = \frac{\hbar}{2m} \Delta\rho - \mathrm{div}(\rho b_+). \qquad (4.5)$$

Thus, if we construct a diffusion process having $b_+(x,t)$ as drift and $\rho(x,0) = |\psi_0(x)|^2$ as initial probability density, i.e.

$$dX_t = b_+(X_t,t)dt + dW_t$$
$$P[X_0 \in A] = \int_A \rho(x,0)dx \qquad (4.6)$$

for any Borel set $A \subset \mathbb{R}^d$, where W_t is a d-dimensional Brownian motion with covariance $\frac{\hbar}{m}t\ 1$, then since the Fokker-Planck equation (4.5) has a unique solution for b_+ and ρ sufficiently smooth, we know that the probability density of the diffusion process X_t is just $\rho(x,t) = |\psi(x,t)|^2$.

From this point of view, the probabilistic interpretation of Schrödinger's equation is very natural. But unfortunately all the above arguments are purely formal, since we don't know whether the solution of (4.5) exists: indeed, $b_+(x,t)$ has no meaning if $\psi(x,t) = 0$. All the classical theorems about the existence of solutions for stochastic differential equations could not be used directly in this case.

On the other hand, physical intuition leads to the conjecture that the sample paths of the diffusion process do not get trapped by the nodal surface of the density (see Section III.4 for a heuristic argument and Section III.5).

This chapter is devoted to the problem of constructing the diffusion process with nice Brownian part and singular drifts which are needed for stochastic mechanics. Using a purely probabilistic approach in terms of suitably defined stopping times as well as some physical and geometrical ideas, the construction of diffusion with singular drifts can be carried out with relative simplicity. This method was used by [15] and inspired by [17a] and [90e].

In the last section of this chapter, we will discuss briefly some previous work in this field.

Let us first describe the problem we have to solve. Suppose we are given at each time $t \in \mathbb{R}_+ = [0,\infty)$ a probability density $\rho(\cdot,t)$ in $\mathbb{R}^d$. Define

$$U = \{(x,t) \in \mathbb{R}^d \times \mathbb{R}_+ \mid \rho(x,t) > 0\} \tag{4.7}$$

and the nodal set

$$N = U^c = \{(x,t) \in \mathbb{R}^d \times \mathbb{R}_+ \mid \rho(x,t) = 0\}. \tag{4.8}$$

Let $b_+ : U \to \mathbb{R}^d$ be given and suppose that ρ and b_+ are related by the forward Fokker-Planck equation on U

$$\partial_t \rho = -\mathrm{div}(\rho b_+) + \nu\Delta\rho \tag{4.9}$$

$\nu = \frac{\sigma^2}{2}$ being the diffusion coefficient. This equation is meaningful if ρ and b_+ satisfy regularity conditions. However, we are only considering a drift b_+ which is defined on U and can therefore be very singular near the boundary of U. The problem consists in constructing a diffusion process X_t with drift b_+ and probability density ρ. In other words, we have to investigate existence and uniqueness of global solutions of the stochastic differential equation

$$dX_t = b_+(X_t,t)dt + dW_t \tag{4.10}$$

where W_t is a standard Wiener process with covariance $2\nu t \mathbb{1}$. If such a process exists its mean forward derivative is equal to b_+ and, moreover, for all bounded measurable functions f and all Borel sets A we have

$$P\{f(X_t) \in A\} = \int_A dx\ \rho(x,t)f(x). \tag{4.11}$$

IV.2 Existence of Nelson's Diffusion Processes

IV.2a Heuristics

Before going through the mathematical discussion, it is worthwhile to get some physical intuition about what might prevent a diffusion to be defined globally and why this is not so.

Let us assume for the moment that (4.6) has a local solution, i.e. the sample paths of the process X_t are defined at least for some finite time interval. What may prevent a trajectory from being defined for all times? This can occur only if the trajectory approaches the nodal set N within finite time and then the drift b_+ is undefined (Fig. 1.a) or if the path escapes to infinity within finite time (Fig. 1.b).

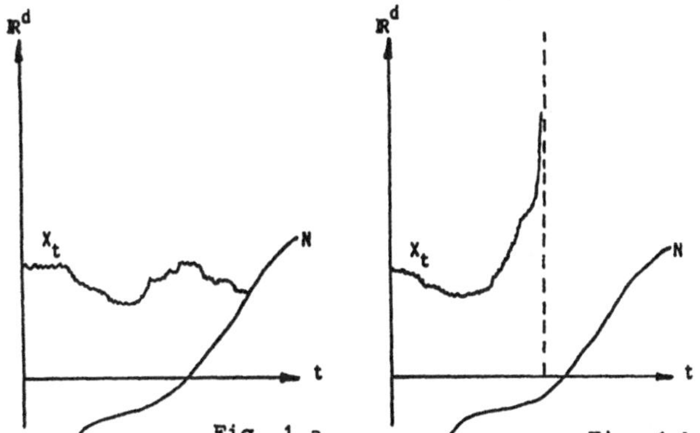

Fig. 1.a   Fig. 1.b

From a physical point of view the unattainability of the nodal set N seems rather plausible. Indeed, in the classical theory of diffusions it is not the drift b_+ itself which has a physical interpretation but the current ρb_+ and even if b_+ is singular on N the product ρb_+ may stay finite. Physically speaking this means that almost no diffusing particles run into the nodes. Moreover, there is another mechanism which can be interpreted as an indication that the particles never reach the nodes: since the singular drift field points away from the node, it will produce a repulsion strong enough to prevent the particle from reaching the nodal surface N.

To prevent explosions (i.e. the escape at infinity) one needs an additional condition (cf. Theorem 4.3).

IV.2b Unattainability of the Nodes and Global Existence

In the sequel, we assume that

$$\text{(A.1)} \quad \rho \in C^{2,1}(\mathbb{R}^d \times \mathbb{R}_+)$$

$$\text{(A.2)} \quad b_+ \in C^{1,0}(U)$$

where $C^{a,b}$ denotes the space of functions with continuous derivative of order a in the space variables and of order b in time variable. Moreover, ρ and b_+ satisfy the Fokker-Planck equation (4.9).

Locally the drift field b_+ is rather regular and the standard theorems grant a unique local solution $(X_t)_{0\leq t\leq \tau}$ of (4.10) up to the stopping time

$$\tau = \inf\{t \geq 0 \mid X_t \notin U\} \tag{4.12}$$

In fact, existence and uniqueness are in the strong sense. We will need the following lemma [116d]

<u>Lemma 4.1</u>: Let $f(x,t)$ be a non-negative function and let τ be a random variable satisfying $0 \leq \tau \leq \zeta \wedge k$ for some $k > 0$. Then

$$\mathbb{E}\left[\int_0^\tau dt\ f(X_t,t)\right] \leq \int_{\mathbb{R}^d} dx \int_0^k dt\ \rho(x,t)f(x,t). \tag{4.13}$$

■

It is convenient to introduce the current velocity v

$$v = \begin{cases} b_+ - \nu\nabla\log\rho & \text{on } U \\ 0 & \text{on } N \end{cases} . \tag{4.14}$$

Moreover, we set $Y_t = (X_t,t)$. We are now ready to prove that the process $(X_t)_{0\leq t\leq\tau}$ does not hit the nodal set N within finite time.

■

<u>Theorem 4.2</u>: Let the process X_t, $0 \leq t < \tau$ be associated to ρ via the Fokker-Planck equation and a drift b_+ such that (A.1) and (A.2) are satisfied. Suppose also that

(A.3) $\rho\,\mathrm{div}\,v \in L^1_{loc}(\mathbb{R}^d \times \mathbb{R}_+)$.

Then for any compact set $K \subset \mathbb{R}^d \times \mathbb{R}_+$

$$P\{\omega \in \Omega \mid \inf_{t<\tau\wedge k} \delta(Y_t, K \cap N) > 0\} = 1 \tag{4.15}$$

δ denoting the Euclidean distance in $\mathbb{R}^d \times \mathbb{R}_+$.

<u>Proof</u>: For $k,\ell \in \mathbb{N}$ we define the compact set

$$K = \{(x,t) \in \mathbb{R}^d \times \mathbb{R}_+ \mid |x| \leq \ell,\ t \leq K\} \tag{4.16}$$

and a subset A of the probability space Ω by

$$A = \{\rho(Y_0) > 0,\ Y_t \in K\ \forall t < \tau \wedge k,\ \inf_{t<\tau\wedge k} \rho(Y_t) = 0\}. \tag{4.17}$$

The set $\{\rho(Y_0) > 0\}$ has probability one. Throughout the proof the continuity of the sample paths $t \to Y_t$ will be essential. A first consequence of this continuity is the fact that in order to prove the theorem it suffices to show that $P(A) = 0 \quad \forall (k,\ell) \in \mathbb{N} \times \mathbb{N}$. Next, we introduce for each $n \in \mathbb{N}$ a stopping time T_n by

$$T_n = \inf\{t < \tau \wedge k \mid Y_t \notin K \text{ or } \rho(Y_t) \le \frac{1}{n}\} \tag{4.18}$$

If $\{t < \tau \wedge k \mid Y_t \notin K \text{ or } \rho(Y_t) \le \frac{1}{n}\}$ is non-void and we set $T_n = \tau \wedge k = k$ otherwise. Since the sequence $(T_n)_{n\in\mathbb{N}}$ is non-decreasing, we define another stopping time $T = \lim_n T_n$. Clearly, $T > 0$ if and only if $|Y_0| \le \ell$ and $\rho(Y_0) > 0$. Thus, we may rewrite the set A in terms of stopping times

$$A = \{T > 0\,,\ \rho(Y_{T_n}) \le \frac{1}{n} \quad \forall n \in \mathbb{N}\}. \tag{4.19}$$

We shall now look at those trajectories for which the density $\rho(Y_t)$ becomes arbitrarily small or, in other words, $\log \rho(Y_t)$ is unbounded from below. An application of Itô's lemma yields

$$\begin{aligned} I_{\{T>0\}} \log \rho(Y_{T_n}) = I_{\{T>0\}} \log \rho(Y_0) + \int_0^{T_n} dt \left[\frac{\partial_t \rho}{\rho} + b_+ \cdot \frac{\nabla\rho}{\rho} + \nu\Delta\log\rho\right](Y_t) \\ + \int_0^{T_n} dW_t \cdot \frac{\nabla\rho}{\rho}(Y_t) \qquad n \in \mathbb{N} \end{aligned} \tag{4.20}$$

$I_{\{\ldots\}}$ being the characteristic function. Now, the expectation of each term in the above formula is to be analyzed.

If $T_n > 0$ and $t \in [0,T_n]$ then $Y_t \in K$ and $\nabla\rho(Y_t)$ is bounded on K by continuity. Moreover, from $T_n > 0$ follows that $\rho(Y_t) \ge \frac{1}{n}$ for $t \in [0,T_n]$. Therefore, the stochastic integral on the right hand side of (4.20) is a martingale (indexed by n) of mean zero. Since $\{T > 0\} \subset \{|Y_0| \le \ell\}$ it follows that

$$\mathbb{E}[|I_{\{T>0\}}\log\rho(Y_0)|] \le \mathbb{E}[I_{\{|Y_0|\le\ell\}} |\log \rho(Y_0)|]$$

$$= \int dx\ \rho(x,0)\ I_{\{|x|\le\ell\}}\ |\log\rho(x,0)|\ . \tag{4.21}$$

The last term in (4.11) is finite since ρ is locally bounded by (A.1) and hence $\mathbb{E}[I_{\{T>0\}}\log\rho(Y_0)]$ exists and is finite.

Let us now consider the left hand side of (4.21). If $T > 0$ then $Y_{T_n} \in K$, $n \in \mathbb{N}$, and $I_{\{T>0\}} \log \rho(Y_{T_n})$ is bounded from above uniformly in $\omega \in \Omega$ and $n \in \mathbb{N}$. Hence $\mathbb{E}[I_{\{T>0\}} \log \rho(Y_{T_n})]$ is uniformly bounded from above.

Assume now that $P(A) > 0$. Our aim is to deduce a contradiction from this assumption. We will denote by $f = f^+ - f^-$ the decomposition of a real-valued function f into its positive and negative parts. Since $A \subset \{T > 0\}$ and $\log \rho(Y_{T_n}) \leq -\log n$ on A it follows

$$\begin{aligned} \mathbb{E}[I_{\{T>0\}} (\log\rho)^-(Y_{T_n})] &\geq [I_A(\log\rho)^-(Y_{T_n})] \\ &\geq P(A) \log n \ . \end{aligned} \tag{4.22}$$

As $\mathbb{E}[I_{\{T>0\}} \log (Y_{T_n})]$ is uniformly bounded from above, this implies

$$\lim_{n\to+\infty} \mathbb{E}\,[I_{\{T>0\}} \log \rho(Y_{T_n})] = -\infty \ , \tag{4.23}$$

and therefore

$$\lim_{n\to+\infty} \mathbb{E}\left[\int_0^{T_n} dt \left\{\frac{\partial_t\rho}{\rho} + b_+ \cdot \frac{\nabla\rho}{\rho} + \nu\Delta\log\rho\right\}(Y_t)\right] = -\infty . \tag{4.24}$$

On the other hand, by reformulating the integrand by means of the Fokker-Planck equation (4.5) we obtain on U

$$\frac{\partial_t\rho}{\rho} + b_+ \cdot \frac{\nabla\rho}{\rho} = \nu\,\frac{\Delta\rho}{\rho} - \operatorname{div} b_+$$

which implies

$$\frac{\partial_t\rho}{\rho} + b_+ \cdot \frac{\nabla\rho}{\rho} + \nu\,\Delta\log \ = \nu\,\frac{\Delta\rho}{\rho} - \operatorname{div} v \ .$$

If $T_n > 0$ and $t \in [0,T_n]$ then $Y_t \in K$ and hence

$$\begin{aligned} &\mathbb{E}\left[\int_0^{T_n} dt \left[\frac{\partial_t\rho}{\rho} + b_+ \cdot \frac{\nabla\rho}{\rho} + \nu\,\Delta\log\rho\right]^- (Y_t)\right] \\ &\quad = \mathbb{E}\left[\int_0^{T_n} dt\ I_{\{T_n\geq 0\}} \left[\nu\frac{\Delta\rho}{\rho} - \operatorname{div} v\right]^- (Y_t)\right] \\ &\quad \leq \mathbb{E}\left[\int_0^{T_n} dt\ I_{\{Y_t\in K\}} \left[\nu\frac{\Delta\rho}{\rho} - \operatorname{div} v\right]^- (Y_t)\right] \\ &\quad \leq \int_{\mathbb{R}^d} dx \int_{\mathbb{R}^+} dt\ I_K(x,t)\ \ [\nu\,\Delta\rho - \rho\,\operatorname{div} v]^-(x,t) \end{aligned}$$

$$\leq \nu \iint_K dx\, dt\, |\Delta\rho(x,t)| + \iint_K dx\, dt\, \rho\, |\mathrm{div}(x,t)|$$

where the last but one step follows from Lemma 1. Since K is compact, the regularity of ρ implies that $\iint_K ds\, dt |\Delta\rho(x,t)| < +\infty$, whereas the finiteness of the second term in the last formula follows from (A.3). Thus we have obtained a uniform upper bound for the expectation $\mathbb{E}\left[\int_0^{T_n} dt\, [\frac{\partial_t \rho}{\rho} + b_+ \cdot \frac{\nabla\rho}{\rho} + \nu\,\Delta \log \rho]\,(Y_t)\right]$ and this contradicts (4.24).

As a result we conclude that A cannot have a non-vanishing probability, i.e. $P[A] = 0$ is established.

Remark 1: For a symmetric diffusion process $v \equiv 0$ and (A.3) is satisfied.

Remark 2: The regularity condition for ρ and b_+ are not be chosen optimally and as a reward for this we get transparent proofs. ∎

In the next theorem, it will be proven that explosions in finite time cannot occur; in other words, the process $(X_t)_{0 \leq t < \zeta}$ cannot disappear at infinity within finite time. This follows essentially from the following assumption on the probability density ρ :

(A.4) There is no continuous path $t \to X_t$ defined on some finite time interval $[0,s)$ such that

$$\sup_{t<s} |X_t| = +\infty \quad \text{and} \quad \inf_{t<s} \rho(X_t,t) > 0 .$$

In addition, we have to strengthen the assumption on ρ and its derivatives from a spatially local to a global one. Then this yields a unique global construction of the diffusion processes. Here both existence and uniquencess are meant in the strong sense.

Theorem 4.3: Let the process X_t , $0 \leq t < \zeta$, ζ being the stopping time defined in (4.12), have a probability density ρ and a drift b_+ such that Assumptions (A.1), (A.2) and (A.4) are satisfied. Suppose also that for all $k \in \mathbb{N}$

(A.5) $\Delta\rho(\cdot,k) \in C_b(\mathbb{R}^d)$

(A.6) $\int dx |\Delta\rho| \in L^1_{loc}(\mathbb{R}_+)$

(A.7) $\int dx\, \rho |\mathrm{div}\, v| \in L^1_{loc}(\mathbb{R}_+)$.

Then $P[\zeta = +\infty] = 1$.

Proof: As in the previous theorem we will use in the following proof suitably defined stopping times. For $k, \ell, n \in \mathbb{N}$ we introduce the stopping times:

$$S^{(1)} = \begin{cases} 0 & \text{if} \quad |X_0| > \ell \\ \zeta \wedge k & \text{if} \quad |X_0| \le \ell \end{cases} \tag{4.25}$$

$$S_n^{(2)} = \inf \{ t < \zeta \wedge k \mid \rho(Y_t) \le \tfrac{1}{n} \} \tag{4.26}$$

if $\{t < \zeta \wedge k \mid \rho(Y_t) \le \frac{1}{n}\}$ is non-void and we set $S_n^{(2)} = \zeta \wedge k = k$ otherwise. We also define the non-increasing sequence

$$S_n = S^{(1)} \wedge S_n^{(2)} \tag{4.27}$$

which gives rise to the limiting stopping time

$$S = \lim_n S_n = S^{(1)} \wedge \zeta \wedge k \, . \tag{4.28}$$

Clearly $S > 0$ if and only if $|X_0| \le \ell$ and $\rho(Y_0) > 0$.
The theorem will be proved if we could show that

$$P[\zeta > k] = 1 \quad \forall k \in \mathbb{N} . \tag{4.29}$$

But we can rewrite

$$\{\zeta > k\} = \{\inf_{t<\zeta\wedge k} \rho(Y_t) > 0 \} \, . \tag{4.30}$$

Moreover, it is sufficient to consider paths starting within compact sets around the origin. Since $P[\rho(Y_0) > 0] = 1$ it suffices to show that $P(B) = 0$ for all $k, \ell \in \mathbb{N}$ where B is defined by

$$B = \{S > 0, \inf_{t<\zeta\wedge k} \rho(Y_t) = 0\} = \{S > 0, \rho(Y_{S_n}) \le \tfrac{1}{n} \ \forall n \in \mathbb{N}\}. \tag{4.31}$$

Using Itô's lemma we obtain

$$I_{\{S>0\}} \log\rho(Y_{S_n}) = I_{\{S>0\}} \log\rho(Y_0) + \int_0^{S_n} dt \left[\frac{\partial_t \rho}{\rho} + b \cdot \frac{\nabla\rho}{\rho} + \nu\Delta\log\rho\right](Y_t)$$

$$+ \int_0^{S_n} dW_t \cdot \frac{\nabla\rho}{\rho} (Y_t) \quad n \in \mathbb{N} . \tag{4.32}$$

If $S_n > 0$ and $t \in [0, S_n]$ then $\rho(Y_t) \ge \frac{1}{n}$ and it follows from (A.5) $E\left[\int_0^{S_n} dt \left|\frac{\nabla\rho}{\rho} (Y_t)\right|^2\right] < +\infty$ and therefore the stochastic integral

$\int_0^{S_n} dW_t \frac{\nabla\rho}{\rho}(Y_t)$ is a martingale of mean zero.

Since $\{S > 0\} \subset \{|X_o| \le \ell\}$ we can write

$$\mathbb{E}\Big[\,|I_{\{S>0\}}\log\rho(Y_o)|\,\Big] \le E\Big[I_{\{|X_o|\le\ell\}}|\log\rho(Y_o)|\Big] = \int_{|x|\le\ell} dx\,\rho(x,0)\,|\log\rho(x,0)|. \tag{4.33}$$

By (A.1) the density ρ is locally bounded and hence $\mathbb{E}[I_{\{S>0\}} \log (Y_o)]$ exists and is finite. To discuss the left hand side of (4.32), we will use the following decomposition

$$1 = I_{\{S_n=0\}} + I_{\{0<S_n<\zeta\wedge k\}} + I_{\{S_n=\zeta\wedge k\}}$$

$$= I_{\{S_n=0\}} + I_{\{0<S_n<\zeta\wedge k\}} + I_{\{S_n=k\}} \cdot$$

Then

$$\mathbb{E}[I_{\{S>0\}\cap\{S_n=0\}}(\log\rho)^+(Y_{S_n})] \le \mathbb{E}[I_{\{|x_o|\le\ell\}}(\log\rho\;)^+(Y_o)]$$

$$= \int_{|x|\le\ell} dx\;\rho(x,0)(\log\rho)^+(x,0), \tag{4.34}$$

$$\mathbb{E}[I_{\{S>0\}\cap\{0<S_n<\zeta\wedge k\}}(\log\rho)^+(Y_{S_n})] = \mathbb{E}[I_{\{0<S_n<\zeta\wedge k}(\log\rho)^+(Y_{S_n})] = 0 \tag{4.35}$$

$$\mathbb{E}[I_{\{S>0\}\cap\{S_n=k\}}(\log\rho)^+(Y_{S_n})] = \mathbb{E}[I_{\{S_n=k\}}(\log\rho)^+(Y_k)]$$

$$< \sup_{x\in\mathbb{R}^d} (\log\rho)^+(x,k). \tag{4.36}$$

Since $\rho(\cdot,k) \in C^1 \cap L^1$ and $\nabla\rho(\cdot,k) \in C_b$ (by (A.5)) it follows that $\rho(\cdot,k) \in C_b$. Thus the decomposition of $\mathbb{E}[I_{\{S>0\}}\log\rho)^+(Y_{S_n}]$ has led to three finite uniform bounds and hence $\mathbb{E}[I_{\{S>0\}}\log\rho(Y_{S_n})]$ is uniformly bounded from above.

Let us now assume that $P(B) > 0$. Since $B \subset \{S > 0\}$ and $\log\rho(Y_{S_n}) \le -\log n$ on B it follows that

$$\mathbb{E}[I_{\{S>0\}}(\log\rho)^-(Y_{S_n})] \ge \mathbb{E}[I_B(\log\rho)^-(Y_{S_n})] \ge P(B)\log n$$

and this implies

$$\lim_{n\to\infty} \mathbb{E}[I_{\{S>0\}} \log\rho(Y_{S_n})] = -\infty, \tag{4.38}$$

$$\lim_{n\to\infty} \mathbb{E}\Big[\int_0^{S_n} dt \left[\frac{\partial_t\rho}{\rho} + b_+ \cdot \frac{\nabla\rho}{\rho} + \nu\Delta\log\rho\right](Y_t)\Big] = -\infty. \tag{4.39}$$

Using the Fokker-Planck equation we obtain then

$$\mathbb{E}\Big[\int_0^{S_n} dt \left[\frac{\partial_t\rho}{\rho} + b_+ \cdot \frac{\nabla\rho}{\rho} + \nu\Delta\log\rho\right](Y_t)\Big] \le \int_{\mathbb{R}^d} dx \int_0^k dt[\nu\Delta\rho - \operatorname{div} v]^-(x,t)$$

$$\le \nu\int_{\mathbb{R}^d} dx \int_0^k dt \; |\Delta\rho(x,t)| + \int_{\mathbb{R}^d} dx \int_0^k dt\rho(x,t) \operatorname{div} v \; (x,t) \tag{4.40}$$

and the two terms are finite by (A.6) and (A.7). This yields a contradiction and we conclude that $P(B) = 0$.

IV.3 Application to Stochastic Mechanics

In stochastic mechanics the probability density $\rho(x,t)$ is related to the solution of the Schrödinger equation by

$$\rho(x,t) = |\psi(x,t)|^2 . \tag{4.41}$$

Moreover, the diffusion constant ν is equal to $\frac{\hbar}{2m}$. We have therefore

$$U = \{(x,t) \in \mathbb{R}^d \times \mathbb{R}_+ \mid \psi(x,t) \neq 0\} \tag{4.42}$$

and b_+ on U is given by (4.2a). Our aim is now to express the conditions (A.3) and (A.7) in terms of the wave function.

Theorem 4.4: Let the wave function ψ be such that

(A.8) $\quad \psi \in C^{2,1}(\mathbb{R}^d \times \mathbb{R}_+)$

(A.9) $\quad \nabla\psi \in L^2_{loc}(\mathbb{R}^d \times \mathbb{R}_+)$.

Then conditions (A.1), (A.2), (A.3) of Theorem 4.2 are satisfied.

Proof: (A.8) implies (A.1) and (A.2). On U let us introduce a real-valued function S (the phase of the wave function ψ) by

$$\psi = \rho^{1/2} e^{iS} \tag{4.43}$$

$$\nabla\log\psi = \frac{1}{2} \nabla\log\rho + i \nabla S \tag{4.44}$$

from which it follows that on U the current velocity takes the form $v = 2\nu\nabla S$. Moreover, we have

$$|\nabla\psi|^2 = \rho\,\frac{1}{4}[(\nabla\log\psi)^2 + (\nabla S)^2] \geq -\nabla\rho\cdot\nabla S = -\frac{1}{2\nu}\,v\cdot\nabla\rho\ . \tag{4.45}$$

Denoting by ψ^* the complex conjugate of ψ we have

$$\rho v = \mathrm{Im}(\psi^*\nabla\psi). \tag{4.46}$$

Condition (A.3) follows then by integrating and using Green's identity. Indeed, the boundary term remains finite by (A.8) and (4.46).

Remark: The conditions we impose are, of course, stronger than necessary because the proof of Theorem 4.3 depends only on the negative part of $\nu\Delta\rho-\rho\nabla\cdot v$. Note that (A.9) also implies $\|\nabla\psi\|_2^2 \in L^1_{loc}(\mathbb{R}^+)$ which is just the finite action (A.12) defined by Carlen and Zheng.

Although the assumption of Theorem 4.4 will be true in many quantum mechanical situations, it would be nice to have conditions in terms of the potential V and the initial wave function $\psi_0(x) = \psi(x,0)$. Let us now state some theorems which give sufficient conditons (see [15] and references therein). For the stationary case, where $\psi(x,t) = e^{-i\frac{Et}{\hbar}}\phi(x)$ we have the following theorem

Theorem 4.5: Let ϕ be a weak solution of

$$(-\frac{\hbar^2}{2m}\,\Delta + V)\phi = E\phi$$

where V is a measurable function and E the eigenvalue. If $V \in C^m(\mathbb{R}^d)$, then $\phi \in C^{m-[d/2]+1}(\mathbb{R}^d)$; here $[d/2]$ denotes the integer part of $d/2$.

So, in the three-dimensional case ψ has at least the same regularity as the potential. In particular $V \in C^2(\mathbb{R}^3)$, then (A.8) is satisfied.

In the case of the time-dependent situation we have

Theorem 4.6: Let $m \in \mathbb{N}$ and suppose that

(A.10) $\psi_0 \in H^{2m}(\mathbb{R}^d)$

(A.11) V and its derivatives up to order $2m-2$ are continuous and bounded in $\mathbb{R}^d$

then (a version of) the solution $\psi_t = e^{-it(-\frac{\hbar^2}{2m}\Delta + V)}\psi_o$ of the Schrödinger equation satisfies $\psi_t \in \bigcap_{\ell=0}^{m} C^{\ell}(\mathbb{R}_+, H^{2(m-\ell)}(\mathbb{R}^d))$.

As a consequence

Theorem 4.7: If conditions (A.10) and (A.11) are satisfied for $2m \geq [\frac{d}{2}] + 3$ then (A.6) and (A.7) hold.

In particular, in the three-dimensional case m has to be greater than two; i.e. we need that $\psi_o \in H^4(\mathbb{R}^3)$ and that V and its derivatives up to order 2 are continuous and bounded.

Theorems 6 and 7 give conditions under which the stochastic mechanical diffusion exists, although the boundedness condition on the potential is physically unsatisfactory. In a given situation, Assumptions (A.10), (A.11) may not hold but (A.6) and (A.7) may be true nevertheless.

IV.4 Alternative Methods to Construct Singular Diffusions

The results exposed in this chapter show that a wide class of diffusion processes with singular drifts can be constructed and that both global existence and uniqueness are in the strong sense. Let us first discuss the results obtained. Conditions similar to (A.3) and (A.7) are tantamount to all constructions of diffusions with singular drifts and from a physical point of view they are not unreasonable. On the other hand, it does not seem to lie within the framework of the method to relax the smoothness conditions (A.1) and (A.2) considerably. As a slight generalization we can replace the requirement for a derivative to exist by Lipschitz condition, which will give uniform bounds, too. As a last remark it should be pointed out that the method carries over to the case where the state space $\mathbb{R}^d$ is replaced by a Riemannian manifold.

There has been some previous work in this field. The stationary case was first considered by Albeverio and Høegh-Krohn [7a] and then by Carmona [23], Nagasawa [89], Albeverio, Fukushima, Karwoski and Streit [6]. The analysis in [6] and [23] is based on the theory of Dirichlet forms and works under mild regularity properties of the probability density, which can be discontinuous (see also Chapter III). In [89] Nagasawa has shown, using probabilistic methods (Dynkin formula), that the diffusion process does not cross the nodal surface of the equilibrium distribution $\rho(x) = |\psi(x)|^2$, where ψ is a solution of a Schrödinger-like equation. In a paper by Blanchard and Zheng [17a] the stationary case was dealt with by using a pathwise conservation law.

The non-stationary situation was solved only recently. In the case where the configuration space is a compact manifold, this was accomplished by Nelson [90e], But the compactness condition can be dropped. For an Euclidean configuration space $\mathbb{R}^d$ Carlen [22a,b,d] established weak existence of solutions of stochastic differential equations with singular drifts. Carlen approaches the problem from a rather analytic point of view. The hard part of his method is to solve a parabolic partial differential equation and to obtain the fundamental solution $p_*(y,t;x,s)$ of this equation. Using p_* and ρ, Carlen constructs a probability measure on $\Omega = (\dot{\mathbb{R}}^d)^{\mathbb{R}_+}$, $\dot{\mathbb{R}}^d$ denoting the one point compactification of $\mathbb{R}^d$, in a standard way. Having this measure, it remains to check that under it the stochastic process is indeed a diffusion with the right coefficients. This last step requires some work since one has only a weak fundamental solution. Nonetheless, Carlen's method produces a honest diffusion. Guerra [60c] has shown that by a slight generalization of Carlen's treatment it is possible to introduce a new class of diffusion processes having very singular drifts. The strategy leading to this class consists of the following steps: Guerra starts from the class of regular diffusions on a bounded time interval; then he introduces a metric on this class and takes the completion with respect to this metric. He proves that in the limit we still get some diffusion processes, even though, in the completion procedure, drifts may become very singular.

In a different way Zheng [115] used tightness results for semimartingales to discuss the diffusions on a Riemannian manifold. The main idea to construct diffusions with a singular drift consists in reducing the problem to the situation where the density ρ is everywhere strictly positive. Both Carlen and Zheng need a kind of global finite action condition to construct the diffusion. In fact, Meyer and Zheng showed [116] that also a local condition guarantees that the diffusions do not reach the nodal set in finite time.

Let us conclude this Section by some remarks.

<u>Remarks</u>: i) Assumptions (A.5) and (A.7) are to be compared to the finite action condition of Carlen [22b] and Zheng [115b]

$$\text{(A.12)} \quad \int dx\, \rho(u^2+v^2) \in L^1_{loc}(\mathbb{R}^+).$$

Strictly speaking, there is no inclusion relation between these conditions, though in a loose sense (A.12) implies (A.6) and (A.7). This can be seen as follows.

Provided that no surface term turns up we have

$$\int_U dx\ \rho\left(\frac{\nabla\rho}{\rho}\right)^2 = -\int_U dx\ \rho\nabla\cdot\left(\frac{\nabla\rho}{\rho}\right). \tag{4.47}$$

Thus the conditions $\int dx\ \rho u^2 \in L^1_{loc}(\mathbb{R}_+)$ does not quite imply $\int_U dx\ \rho|\nabla\cdot(\frac{\nabla\rho}{\rho})| \in L^1_{loc}(\mathbb{R}_+)$, which were sufficient for (A.6), since on U

$$\Delta\rho = \rho\left[\left(\frac{\nabla\rho}{\rho}\right)^2 + \nabla\cdot\left(\frac{\nabla\rho}{\rho}\right)\right] \tag{4.48}$$

$$|\Delta\rho| = \rho\left[\left(\frac{\nabla\rho}{\rho}\right)^2 + \left|\nabla\cdot\left(\frac{\nabla\rho}{\rho}\right)\right|\right]. \tag{4.49}$$

Similarly, $2|uv| \le u^2 + v^2$ and thus

$$\int dx\ \rho|uv| \le \frac{1}{2}\int dx\ \rho(u^2 + v^2). \tag{4.50}$$

Therefore (provided there is no surface contribution),

$$\int dx\ \rho \operatorname{div} v = \frac{1}{2}\int dx\ \rho u \tag{4.51}$$

is an $L^1_{loc}(\mathbb{R}_+)$ by virtue of (A.12). However, (A.7) require a $|\nabla.v|$-term rather than merely $\nabla\cdot u$. In conclusion, (A.6) and (A.7) constitute a different kind of finite action condition than that of Carlen and Zheng, but often they will be a consequence of (A.12).

ii) The basic strategy in our proofs is to find appropriate estimates on $\log\rho$, and this is done by means of Itô's lemma. In this respect, our method is similar to the one employed by Nelson [90c]. His proof does not require the finite action condition (A.12) but works with the weaker condition $\int_0^k dt \int dx\ \rho|u\cdot v| < \infty$ which, according to (4.51), is related to (A.7).

iii) In [6, Th. 4.2] the unattainability of nodal set is proved under the assumption that, in perpendicular direction to the nodes, the probability density falls off to zero sufficiently fast (essentially, faster than linearly). In a way this corresponds to (A.6), although our assumption has the drawback of not distinguishing between perpendicular and tangential properties of the density.

V. STOCHASTIC VARIATIONAL PRINCIPLES

V.0 Introduction

Before proceeding to a detailed examination of stochastic variational principles, it is probably desirable that we review the situation in classical mechanics and that we make a few general remarks. The calculus of variation has been useful as a unifying principle in mechanics and, more generally, as a guide for the determination of new laws of physics. One of the most widely applicable variational statements of classical mechanics is known as Hamilton's principle; it states that the trajectories of many classical dynamical systems are the solutions of some variational problem involving an energy integral. The equations of Euler-Lagrange and Hamilton do not represent new theories, but only new ways of looking at dynamics, and the resulting equations of motion are the same as those derived by Newton's law.

The calculus of variation is concerned with the following problem: In a given set of admissible functions find a function of a given functional for which the latter is an extremum with respect to all functions of the domain. In a concrete problem, the first problem we have to discuss is also the determination of the class of admissible functions.

Let us return to stochastic diffusion equations

$$dX_t = b_+(X_t,t)dt + dW_t \ .$$

If $X_t^{(1)}$ and $X_t^{(2)}$ are diffusion processes

$$X_t^{(i)} = X_o^{(i)} + \int_o^t b_+{}^{(i)}(X_\tau^{(i)},\tau)d\tau + W_t \qquad i = 1,2$$

let us remark that the sum $X_t^{(1)} + X_t^{(2)}$ is no longer a diffusion process. Therefore, we must enlarge the class of processes we consider in view of formulating stochastic variational principles which are relevant for stochastic mechanics. This leads us to the consideration of some classes of semimartingales.

V.1 The Classes $S(P)$ and $S(P,F)$

On a bounded interval $I = [0,T] \subset \mathbb{R}_+$ let us consider a family

$(P_t)_{t\in I}$ of σ-algebras which is increasing, continuous from the right and, moreover, such that P_o contains all the P-negligible sets. The third condition ensures that every P_t is P-complete (see Appendix).

We recall that a process X_t is a $(P_t)_{t\in I}$ - continuous semimartingale if X_t admits the canonical decomposition

$$X_t = X_o + M_t + A_t \tag{5.1}$$

such that

(i) X_o is a P_o-measurable random variable.

(ii) M_t is a $(P_t)_{t\in I}$ - local martingale, i.e. there exists a sequence $(T_n)_{n\in\mathbb{N}}$ of stopping times T_n, $T_n \uparrow \infty$ such that $M_{T_n \wedge t}$ is a $(P_t)_{t\in I}$ - martingale for every fixed n, $M_o = 0$ a.s.

(iii) A_t is an adapted process of bounded variation, i.e. $A_t \in P_t$ for all $t \in I$ and for almost every fixed $\omega \in \Omega$, $A_t(\omega)$ is a function of bounded variation on every bounded interval of time $[0,t] \subset I$, $A_o = 0$ a.s.

We must now restrict the class of semimartingales we consider in such a way that this class includes some diffusion processes.

Let $S(P)$ be the collection of semimartingales admitting the decomposition $X_t = X_o + M_t + A_t$ such that

i) $X_o \in L^2(P_o)$

ii) M_t is a continuous square-integrable martingale on $[0,a]$ with $M_o = 0$

iii) $A_t = \int_0^t H_s \, ds$, where H_s is a (P_s)-adapted process such that $\mathbb{E}\left[\int_0^T |H_s|^2 \, ds\right] < +\infty$ (finite energy condition).

Let us first remark that the last condition is not satisfied by all diffusion processes. Moreover, the process H_t is the "forward derivative" of X_t in the sense of Nelson (see Sect. II.2). For this reason, we write often $H_t = D_+X_t$.
On $S(P)$ we define a norm $\|\cdot\|_{S(P)}$ by

$$\|X\|^2_{S(P)} \equiv \mathbb{E}\left[|X_T|^2 + \int_0^T |H_s|^2 ds\right] . \tag{5.2}$$

We are now prepared to state our first result.

<u>Lemma 5.1</u>: $(S(P), \|\cdot\|_{S(P)})$ is a Hilbert space.

<u>Proof</u>: By Schwarz's inequality we have

$$\mathbb{E}\left[\left(\int_0^t |H_s| ds\right)^2\right] \le t \ \mathbb{E}\left[\int_0^t |H_s|^2 ds\right] . \tag{5.3}$$

Since $X_T - X_t = \int_t^T H_s ds + M_T - M_t$ it follows

$$X_t + (M_T - M_t) = X_T - \int_t^T H_s ds . \tag{5.4}$$

Using now the fact that the random variables X_t and $(M_T - M_t)$ are orthogonal, we obtain

$$\mathbb{E}[|X_t|^2] + \mathbb{E}[(M_T - M_t)^2] \le 2\,\mathbb{E}\,|X_T|^2] + 2(T-t)\,\mathbb{E}[\int_t^T |H_s|^2 ds] . \tag{5.5}$$

Thus, let $\{X^{(n)}\}_{n\in\mathbb{N}}$ be a Cauchy sequence in $S(P)$, i.e. $\sup_m \|X^{(n)} - X^{(n+m)}\|_{S(P)} \to 0$ for $n \to +\infty$. This implies in particular that $\sup_m \mathbb{E}[|M_T^{(n)} - M_T^{(n+m)}|^2] \to 0$ for $n \to \infty$,
$\sup_m \mathbb{E}[|X_0^{(n)} - X_0^{(n+m)}|^2] \to 0$ for $n \to +\infty$ and
$\sup_m \mathbb{E}[\int_0^T |H_s^{(n)} - H_s^{(n+m)}|^2 ds] \to 0$ for $n \to +\infty$. Since L^2 is complete, there exist $M_T^{(\infty)}$, $X_0^{(\infty)}$ and $H_t^{(\infty)}$ such that

$$\lim_{n\to\infty} \mathbb{E}[|M_T^{(n)} - M_T^{(\infty)}|^2] = \lim_{n\to\infty} \mathbb{E}[|X_0^{(n)} - X_0^{(\infty)}|^2] = 0 \tag{5.6}$$

and
$$\lim_{n\to\infty} \mathbb{E}[\int_0^T |H_s^{(n)} - H_s^{(\infty)}|^2 ds] = 0 . \tag{5.7}$$

Denoting now by $X^{(\infty)}$ the element of $S(P)$ defined by

$$X_t^{(\infty)} = \int_0^t H_s^{(\infty)} ds + M_t^{(\infty)} + X_0^{(\infty)} \tag{5.8}$$

where $M_t^{(\infty)} = \mathbb{E}[M_T^{(\infty)} | P_t]$ it is easy to verify that

$$\lim_{n\to\infty} \|X^{(n)} - X^{(\infty)}\|_{S(P)} = 0 . \tag{5.9}$$

□

Now, given a second filtration F such that F_{T-t} satisfies the same conditions as P_t we say that a continuous process $X \in S(P,F)$ if

$X_t \in S(P_t)$ and $X_{T-t} \in S(F_{T-t})$. Let $X_{T-t} = X_T + \int_0^t K_s ds + \hat{M}_t$. We denote

$$D_- X_t = -K_{T-t} \ . \tag{5.10}$$

On $S(P,F) \subset S(P)$ we introduce the norm $\|\cdot\|_{S(P,F)}$

$$\|X\|_{S(P,F)} = \{[|X_0|^2 + \int_0^T (|D_+X_s|^2 + |D_-X_s|^2)ds]\}^{1/2} \ . \tag{5.11}$$

As in Lemma 1, we can prove that

<u>Lemma 5.2</u>: $(S(P,F), \|\circ\|_{S(P,F)})$ is complete.

<u>Remark</u>: On $S(P,F)$ the following norm

$$\|X\|^2 = E[|X_0|^2 + |X_T|^2 + \int_0^T (|D_+X_s|^2 + |D_-X_s|^2)ds] \tag{5.12}$$

is equivalent to $\|\cdot\|_{S(P,F)}$. ■

V.2 <u>Strongly Convex Functionals</u>

Let $(X, \|\cdot\|)$ be a normed space and let $K \subset X$ be a convex subset. We say that a functional $f: K \subset \mathbb{R}$ is <u>strongly convex</u> if there exists a constant $C > 0$ such that we have

$$(1-\lambda)f(a) + \lambda f(b) - f[(1-\lambda)a + \lambda b] \geq C\lambda(1-\lambda)\|b-a\|^2 \tag{5.13}$$

for all $a,b \in K$ and $\lambda \in [0,1]$. The strongly convex functionals form a subclass of the class of convex functionals.
We are interested in the following problem: Find conditions which assure that a function

$$f : X \to \mathbb{R}$$

attains its minimum. We recall that a function f is called <u>lower-semicontinuous</u> if $\lim_n x_n = x$ implies $\underline{\lim}_n f(x_n) = f(x)$.
We now prove a theorem giving such sufficient conditions:

<u>Theorem 5.3</u>: Let $f: K \to \mathbb{R}$ be a strongly convex and lower-semicontinuous functional defined on a closed convex set $K \subset X$. If f is bounded from below there exists a unique $x_\infty \in K$ such that

$$f(x_\infty) = \inf_{x \in K} f(x) . \tag{5.14}$$

■

<u>Proof</u>: Let $\{X_n\}_{n \in \mathbb{N}}$ be a sequence in K such that

$$\lim_{n\to\infty} f(x_n) = \inf_{x \in K} f(x) > -\infty .$$

Then we have also

$$\lim_{n,m\to\infty} f(\tfrac{1}{2}(x_n + x_m)) = \inf_{x\in K} f(x) .$$

Using now the strong convexity of f for $\lambda = \frac{1}{2}$

$$c\tfrac{1}{4} \|x_n - x_m\|^2 \leq \tfrac{1}{2} f(x_n) + \tfrac{1}{2} f(x_m) - f(\tfrac{1}{2}(x_n + x_m))$$

we conclude that $\{X_n\}_{n\in\mathbb{N}}$ is a Cauchy sequence. Since K is closed, there exists $X_\infty \in K$ such that $\lim_{n\to\infty} X_n = X_\infty$. Moreover, the lower-semicontinuity of f implies

$$f(x_\infty) = \inf_{x\in K} f(x) .$$

To prove the uniqueness, note that if y and x_∞ realize the minimum of f on K, then $\frac{1}{2}(y + x_\infty) \in K$ and

$$f(\tfrac{1}{2}(y + x_\infty)) < \inf_{x\in K} f(x)$$

unless $y = x_\infty$.

■

V.3 <u>The Yasue Action</u>

Let $V(x,t)$ be a potential function. In classical mechanics, one considers the action functional

$$J_c^T = \int_0^T \{\tfrac{m}{2} \dot{x}_t^2 - V(X_t,t)\}dt . \tag{5.15}$$

In stochastic mechanics, we define following Yasue [112] a similar action

$$J^T = \mathbb{E}\left[\int_0^T [\tfrac{m}{4} (|D_+X_s|^2 + |D_-X_s|^2) - V(X_s,s)]ds \right] . \tag{5.16}$$

At this point, it is convenient to make some hypotheses about the potential.

We suppose that the second derivatives of $V(\cdot,t)$ along all straight lines are bounded:

$$\frac{d^2}{d\lambda^2} V(x+\lambda e,t) \le k \quad \forall x \in \mathbb{R}^d \quad \text{and for all unit vectors} \quad e. \tag{5.17}$$

Moreover, we suppose that there exists a constant C such that

$$V(x,t) \le C(1+|x|)^2 \quad \forall x \in \mathbb{R}^d \quad \forall t \in I \ . \tag{5.18}$$

The first basic fact about the functional J^T is presented in the next theorem.

Theorem 5.4: If $\xi \in S(P,F)$ is such that $J^T(\xi) < +\infty$ and if the convex set

$$K_\xi = \{X \in S(P,F) \mid X_o = \xi_o \ , \ X_T = \xi_T\} \tag{5.19}$$

is closed, then the functional J^T is strongly convex on K_ξ if $T^2 < m/k$. As a consequence, there exists a unique element in K_ξ which minimizes J^T.

Proof: Remark first that the class of admissible functions we consider is much the same as for classical mechanics. Let X,Y be two elements of K_ξ and denoted by $Z = X-Y$. Then $Z_o = Z_T = 0$. Define functionals

$$J_1^T(X) = \mathbb{E}\,[\tfrac{m}{4}\int_o^T (|D_+X_s|^2 + |D_-X_s|^2)ds] \tag{5.20}$$

$$J_2(X) = \mathbb{E}\,[\int_o^T V(X_s,s)ds] \ . \tag{5.21}$$

It is easy to verify that

$$(1-\lambda)J_1^T(X) + \lambda J_1^T(Y) - J_1^T[(1-\lambda)X+\lambda Y] = \lambda(1-\lambda)J_1^T(Z) \tag{5.22}$$

$$= \frac{m}{4}\,\lambda(1-\lambda)\,\|Z\|^2_{S(P,F)}.$$

Using (5.17) we obtain

$$(1-\lambda)V(X_s,s) + \lambda V(Y_s,s) - V[(1-\lambda)X_s+\lambda Y_s,s] \le k\lambda(1-\lambda)|Z_s|^2 \ . \tag{5.23}$$

From (5.5) we see that

$$\int_o^T \mathbb{E}[|Z_t|^2]dt \le \frac{1}{2}\,T^2\,\mathbb{E}\,[\int_o^T |D_+Z_s|^2 ds] \ .$$

Thus, using the similar inequality for $D_- Z_s$, we obtain

$$\int \mathbb{E}[|Z_s|^2]ds \leq \frac{1}{4} T^2 \|Z\|^2_{S(P,F)}$$

and therefore from (5.22)

$$(1-\lambda)J_2^T(X) + \lambda J_2^T(Y) - J_2^T[(1-\lambda)X + \lambda Y] \leq \frac{k}{4} T^2 \lambda(1-\lambda) \|Z\|^2_{S(P,F)} . \qquad (5.24)$$

Hence combining (5.21) and (5.24) it follows

$$(1-\lambda)J^T(X) + \lambda J^T(Y) - J^T[(1-\lambda)X + \lambda Y] \geq \frac{1}{4} (m - kT^2) \|Z\|^2_{S(P,F)}$$

and therefore J^T is strongly convex on the closed convex set $K_\xi \subset S(P,F)$. Hence by Theorem 5.3 there exists a unique element in K_ξ which minimizes J^T . ∎

To discuss the connections between the minimizing element and the solution of the stochastic Newton law, we prove first

Lemma 5.5: Let X,Y be in $S(P,F)$ and suppose that D_+X_t and D_-X_t are also in $S(P,F)$. Then, denoting

$$a(X_t) = \frac{1}{2} (D_+D_- + D_-D_+)X_t \qquad (5.25)$$

we have

$$\mathbb{E}\left[\int_0^T a(X_s).Y_s\, ds\right] = Y_t \cdot \frac{1}{2}(D_+ + D_-)X_t\Big|_0^T - \mathbb{E}\left[\int_0 \frac{1}{2}(D_+X_s.D_+Y_s + D_-X_s.D_-Y_s)ds\right] . \qquad (5.26)$$

Proof: Using the formula of integration by parts (see 2.52) we obtain

$$\mathbb{E}\left[\int_0^T D_-D_+X_s \cdot Y_s ds\right] = Y_t \cdot D_+X_t\Big|_0^T - \mathbb{E}\left[\int_0^T D_+X_s \cdot D_+Y_s ds\right] \qquad (5.27)$$

$$\mathbb{E}\left[\int_0^T D_+D_-X_s \circ Y_s ds\right] = Y_t \cdot D_-X_t\Big|_0^T - \mathbb{E}\left[\int_0^T D_-X_s \circ D_-Y_s ds\right] \qquad (5.28)$$

which implies the result. ∎

Assuming, moreover, that the potential V is such that

$$|V(x+\lambda y,t) - V(x,t) - \lambda \nabla_x V(x,t)\cdot y| \le C\lambda^2(1+|x|^2+|y|^2) \tag{5.29}$$

for all $x,y \in \mathbb{R}^n$ and all λ with $0 < \lambda \le 1$ then it is easy to check the following property:

Let $X,Y \in S(\mathcal{P},\mathcal{F})$ such that $Y_0 = Y_T = 0$ then we have

$$\frac{d}{d\lambda} J^T(X+\lambda Y)\Big|_{\lambda=0} = \frac{m}{2}\,\mathbb{E}\Big[\int_0^T (D_+X_t\cdot D_+Y_t + D_-X_t\cdot D_-Y_t)\,dt\Big] - \mathbb{E}\Big[\int_0^T Y_t\cdot\nabla_x V(X_t,t)\,dt\Big] \;. \tag{5.30}$$

We are now prepared to state our next theorem.

<u>Theorem 5.6 (Yasue)</u>: Suppose $V\colon \mathbb{R}^n \times I \to \mathbb{R}$ satisfies (5.17), (5.18) and (5.28), then for T small enough the unique minimizing point of the action functional J^T is characterized by

$$\frac{m}{2}\,\mathbb{E}\Big[\int_0^T (D_+X_t\cdot D_+Y_t + D_-Y_t\cdot D_-Y_t)\,dt\Big] = \mathbb{E}\Big[\int_0^T \nabla_x V(X_t,t)\cdot Y_t\,dt\Big] \tag{5.31}$$

for all $Y \in S(\mathcal{P},\mathcal{F})$ such that $Y_0 = Y_T = 0$. ∎

Next, we would like to discuss Newton's law in the mean. Theorem 5.6 implies the following

<u>Corollary 5.7</u>: Let V and T be as in Theorem 5.3. Suppose that $X \in S(\mathcal{P},\mathcal{F})$ satisfies Newton's law in the mean

$$m\,a(X_s) = -\nabla_x V(X_s,s) \tag{5.32}$$

and that it is such that D_+X_t , $D_-X_t \in S(\mathcal{P},\mathcal{F})$, then the process X is the unique minimizing point of the Yasue action J^T.

But unfortunately, the converse is not valid. The minimizing extremal point, the existence of which is proved, is not necessarily a solution of the stochastic Newton law. Let us also remark that, moreover, nothing ensures that the minimizing point of J^T in $S(\mathcal{P},\mathcal{F})$ is a diffusion process. These two restrictions make clear that the results we obtain are not very satisfactory from the point of view of stochastic mechanics.

V.4 Construction of Diffusion Processes by a Forward Stochastic Variational Principle

In this section, we discuss a method to construct diffusion processes with constant diffusion coefficient using a stochastic variational principle. We emphasize that the result we obtain is much the same as for classical mechanics in the sense that the minimality of the action is equivalent to the (stochastic) Newton law.

(Ω, P, P) is a given probability space and on a bounded time interval $I = [0,T]$ we consider, as in Sect. V. 1, an increasing filtration $(\mathcal{P}_t)_{t\in I}$ which is continuous from the right and such that $\mathcal{P}_o$ contains all P-null sets. We consider the class $S(\mathcal{P})$ of continuous semimartingales and recall that $(S(\mathcal{P}), \|\cdot\|_S)$ with

$$\|X\|_{S(\mathcal{P})} = \mathbb{E}[\,|X_T|^2 + \int_0^T |H_s|^2\, ds] \tag{5.33}$$

is a Hilbert space (Lemma 5.1).

Now, let $(W_t)_{t\in I}$ be a Brownian motion with initial value W_o and let $(\mathcal{P}_t)$ be a filtration containing $\sigma(W_s | s \leq t)$. We denote by $S_W(\mathcal{P}) \subset S(\mathcal{P})$ the subset of $S(\mathcal{P})$ such that:

$$X \in S_W(\mathcal{P}) \Leftrightarrow \begin{cases} \text{(i) } X_o = W_o \\ \text{(ii) the martingale part of } (X_t) \text{ is } (W_t). \end{cases} \tag{5.34}$$

On $S_W(\mathcal{P})$ we can define a new distance as follows. Let $X_t = \int_0^t H_s ds + W_t$ and $Y_t = \int_0^t K_s ds + W_t$, then we set

$$\|X-Y\|^2 = \mathbb{E}[\int_0^T |H_s - K_s|^2\, ds] . \tag{5.35}$$

To show that on $S_W(\mathcal{P})$ this new distance is equivalent to the distance given by $\|\ \|_S$ we remark that

$$\|X-Y\|^2 \leq \|X-Y\|^2_{S(\mathcal{P})} \leq (T+1)\, \|X-Y\|^2 . \tag{5.36}$$

The next lemma describes elementary properties of $S_W(\mathcal{P})$.

<u>Lemma 5.8</u>: i) $S_W(\mathcal{P})$ is an affine subset of $S(\mathcal{P})$

ii) $S_W(\mathcal{P})$ is complete for the distance $\|\cdot\|$

iii) For $X, Y \in S_W(\mathcal{P})$ we have

$$\mathbb{E}[\int_0^T |X_t - Y_t|^2 dt] \leq \frac{1}{2} T^2\, \|X-Y\|^2 . \tag{5.37}$$

Proof: i) and ii) are a corollary of Lemma 5.1. Let $X,Y \in S_w(P)$, then

$$\mathbb{E}[\int_0^T |X_t - Y_t|^2 dt] = \mathbb{E}[\int_0^T |\int_0^t (H_s - K_s)^2 ds| dt]$$

and from Schwarz's inequality

$$|\int_0^t (H_s - K_s) ds|^2 \le t \int_0^t |H_s - K_s|^2 ds .$$

Then we get

$$\mathbb{E}[\int_0^T |X_t - Y_t|^2 dt] \le \mathbb{E}[\int_0^T t \, dt \int_0^T |H_s - K_s|^2 ds] = \frac{1}{2} T^2 \|X-Y\|^2.$$

Let $V(x,t)$, $x \in \mathbb{R}^n$, $t \in I$ be a potential function. On $S_w(P)$ we consider the action functional J defined in the following way: Let $X_t = \int_0^t H_s ds + W_t$, then

$$J(X) = \mathbb{E}[\int_0^T (\frac{m}{2} |H_t|^2 - V(X_t,t)) dt] . \qquad (5.38) \quad \blacksquare$$

To obtain the strong convexity of the action J, we impose on V exactly the same conditions as in V.1. Using Theorem 2.1, we obtain the following

Theorem 5.9: Let V be such that (5.17) and (5.18) are satisfied. Then, for $T^2 < \frac{m}{k}$ the action J is strongly convex on $S_w(P)$ and there exists a unique element ξ of $S_w(P)$ which minimizes J, i.e.

$$-\infty < J(\xi) = \inf_{X \in S_w(P)} J(X) < +\infty .$$

Proof: The proof of Theorem 5.9 is entirely analogous to the proof of Theorem 5.4, and so it will be left as an exercise (see [17b]). $\quad \blacksquare$

Next we turn to a discussion of the properties of the unique extremal point ξ of J in $S_w(P)$. Let us suppose that V satisfies the same conditions as in V.3.

We are now in position to formulate the first basic theorem of this section:

Theorem 5.10: Suppose that $V: \mathbb{R}^n \times I \to \mathbb{R}$ satisfies the conditions (5.17), (5.18) and (5.29). Let $T^2 < \frac{m}{k}$. An element $X \in S_w(P)$ is the extremal point of J if and only if

$$H_s = \frac{1}{m} \, \mathbb{E}[\int_s^T \nabla_x V(X_u, u) du | P_s]$$

$$= \int_0^s - \frac{1}{m} \nabla_x V(X_u, u) du + \mathbb{E}[\int_0^T \frac{1}{m} \nabla_x V(X_u, u) du | P_s]. \qquad (5.39)$$

Remark: Using the notations of Chap. III, by the remark in Section V.1, (5.39) is just equivalent to

$$\begin{cases} m \, D_+ D_+ X_s = -\nabla_x V(X_s, s) \\[2mm] D_+ X_T = 0 \quad \text{a.s.} \end{cases} \qquad (5.40)$$

Proof: We recall first a classical result. Let $f(t)$ and $g(t)$ be two integrable measurable functions, then for every $0 < T < +\infty$

$$\int_0^T \left(\int_0^t f(s) ds \right) g(t) dt = \int_0^T \int_0^T I_{\{s \le t\}}(s,t) f(s) g(t) ds \, dt$$

$$= \int_0^T \left(\int_s^T g(t) \, dt \right) f(s) ds. \qquad (5.41)$$

Now, let $X_t = \int_0^t H_s ds + W_t$ be the extremal point of J. For any $Y_t = \int_0^t K_s ds + W_t$ in $S_w(P)$ we have

$$J(Y) - J(X) = \mathbb{E}[\int_0^T m \, H_t (K_t - H_t) - \nabla_x V(X_t, t)(Y_t - X_t) dt] + o(\|Y - X\|) \qquad (5.42)$$

Since X is the extremal point of J in $S_w(P)$, we observe that

$$\mathbb{E}[\int_0^T \{m \, H_t (K_t - H_t) - \nabla_x V(X_t, t)(Y_t - X_t)\} dt] = 0 \qquad (5.43)$$

which can be rewritten as

$$\mathbb{E}[\int_0^T \{m \, H_t (K_t - H_t) - \nabla_x V(X_t, t) \cdot \int_0^t (K_s - H_s) ds\} dt] = 0. \qquad (5.44)$$

Using (5.41) we have also

$$\mathbb{E}[\int_0^T \{m \, H_t (K_t - H_t) - (\int_t^T \nabla_x V(X_s, s) ds)(K_t - H_t)\} dt] = 0. \qquad (5.45)$$

Inserting now

$$K_t - H_t = \mathbb{E}[m \, H_t - \int_t^T \nabla_x V(X_s, s) ds | P_t] \qquad (5.46)$$

we obtain

$$m\ H_t = \mathbb{E}\left[\int_t^T \nabla_x\ V(X_s,s)ds \mid P_t\right]\ . \tag{5.47}$$

Conversely, if (5.47) holds we have also ((5.43) and X is the extremal point of J. □

In (5.40) the relation $m\ D_+D_+X_s = -\nabla_x\ V(X_s,s)$ can be interpreted as a stochastic Newton equation. Since the condition $DX_T = 0$ a.s. is too restrictive, we modify a little bit the action functional J.

Let $f \in C^2(\mathbb{R}^d)$ such that $\forall x \in \mathbb{R}$

$$\begin{cases} |f(x)| < C(1+|x|^2) \\ \dfrac{d^2}{d\lambda^2}\ f(x+\lambda e) \leq k \quad \text{for all unit vectors} \quad e \in \mathbb{R}^d \\ |f(x+\lambda y) - f(x) - \nabla f(x)\ y| \leq C\lambda^2(1+|x|^2+|y|^2) \\ \qquad\qquad \forall x,y \in \mathbb{R}^d \quad \forall \lambda \in [0,1]. \end{cases} \tag{5.48}$$

Theorem 5.11: Let $V: \mathbb{R}^d \times I \to \mathbb{R}$ such that (5.17), (5.18) and (5.29) are satisfied and $f \in C^2(\mathbb{R}^d)$ verifies (5.1). Let J_f be the functional on $S_w(P)$ defined by

$$J_f(X) = J(X) + \frac{T}{2}\mathbb{E}[f(X_T)]\ . \tag{5.49}$$

Then, if $T^2 < \frac{m}{2k}$, the functional J_f is strongly convex. Moreover, an element $X \in S_w(P)$ is the unique extremal point of J_f if and only if

$$H_s = \int_0^s -\frac{1}{m}\ \nabla_x\ V(X_u,u)du + \mathbb{E}\left[\int_0^T\left(\frac{1}{m}\ \nabla_x\ V(X_u,u)du + \frac{T}{2}\ \nabla_x\ f(X_T)\right) \mid P_s\right]\ . \tag{5.50}$$

Remark: (5.50) can be put in the form

$$\left.\begin{aligned} mD_+D_+X_s &= -\nabla_x\ V(X_s,s) \\ D_+X_T &= \frac{T}{2}\ \nabla_x\ f(X_T) \end{aligned}\right\} \tag{5.51}$$

Proof: The proof being almost the same as that of Theorem (5.10), we leave the details to the reader.

Now, let us consider the solution X_t of the following stochastic differential equation

$$dX_t = b_+(X_t,t)dt + dW_t \tag{5.52}$$

where W_t has the quadratic variation

$$d\langle W^i,W^j\rangle_t = \nu\delta^{ij}\, dt \tag{5.53}$$

and let f and V be two functions such that

$$\frac{T}{2}\nabla_x f(x) = b_+(x,T) \tag{5.54}$$

$$-\nabla_x V(x,t) = m\left[\frac{\partial b_+}{\partial t}(x,t)+(b_+(x,t)\cdot\nabla_x)b_+(x,t)+\frac{\nu}{2}\Delta_x b_+(x,t)\right]. \tag{5.55}$$

If the hypotheses of Theorem 5.11 are satisfied, then X is the unique extremal point of the action $J_f : S_W(P) \to \mathbb{R}$.

We have proved that the functional J_f has a unique minimizing element ξ in $S_W(P)$

$$\xi_t = \int_0^t H_s\, ds + W_t\ . \tag{5.56}$$

If we can establish, moreover, that ξ is a Markov process, we will be able to conclude that

$$H_s = b_+(\xi_s,s)\ . \tag{5.57}$$

In other words, the process ξ_t will be a diffusion process. We formulate the second basic theorem of this section:

<u>Theorem 5.12</u>: Under the same hypotheses as in Theorem 5.11 the unique extremal point of the action functional $J_f : S_W(P) \to \mathbb{R}$ is a Markov process which is the solution of the following stochastic differential equation

$$d\xi_t = b_+(\xi_t,t)dt + dW_t\ ,\quad \xi_o = W_o \tag{5.58}$$

where b is given by

$$b_+(\xi_t,t) = \mathbb{E}\left[\int_t^T \frac{1}{m}\nabla_x V(\xi_u,u)du + \frac{T}{2}\nabla_x f(\xi_T)\,|\,\xi_t\right]\ . \tag{5.59}$$

<u>Proof</u>: For a proof of this more technical result, see [17b].

Let us emphasize that Theorem 5.12 gives a method of stochastic mechanics to prove the existence of the solution of the stochastic differential equation

$$d\xi_t = b_+(\xi_t,t)dt + dW_t \ .$$

Now, we discuss briefly the relation between the above result and the results obtained in Chap. III. In Chap. III, following E. Nelson, we have considered another stochastic acceleration defined as

$$a(X_t) = \frac{1}{2}(D_+D_- + D_-D_+)X_t \ . \tag{5.60}$$

Let $\psi(x,t) = e^{R(x,t)+i\,S(x,t)}$ be a solution of Schrödinger's equation

$$i\hbar \frac{\partial\psi}{\partial t} = -\frac{\hbar^2}{2m}\Delta\psi(x,t) + V(x,t)\psi(x,t). \tag{5.61}$$

Setting $H(x,t) = \frac{\hbar}{m}\{\nabla_x R(x,t) + \nabla_x S(x,t)\}$ and $\rho(x,t) = |\psi(x,t)|^2$ we know that under some regularity conditions (see Chapter IV), there exists a diffusion process

$$dX_t = H(X_t,t)dt + dW_t \tag{5.62}$$

$$d<W^i,W^j>_t = \frac{\hbar}{m}\delta^{ij}\,dt \tag{5.63}$$

whose probability density at time t is just $\rho(x,t)$ and, moreover, that the following stochastic Newton equation holds

$$m\,a(X_t,t) = -\nabla_x V(X_t,t) \ . \tag{5.64}$$

Now, expressing $a'(X_t,t) = D_+D_+X_t$ as a function of X_t, it can be easily shown that the above equation is formally equivalent to

$$m\,a'(X_t,t) = -\nabla_x[V(X_t,t) - V'(X_t,t)] \tag{5.65}$$

with

$$V'(X,t) = \frac{\hbar^2}{m}\frac{\Delta e^R}{e^R} \ . \tag{5.66}$$

V.5 Other Approaches to Stochastic Calculus of Variations

In this section, we present the main ideas of the various attempts to introduce a calculus of variations for stochastic processes which allows to characterize the dynamics of probabilistic systems as discussed in Chapter III by extremal properties of some non-linear functionals of the process.

The first attempt to introduce variational methods in the framework of Nelson's stochastic mechanics is due to K. Yasue [112] (see also [114a] for a review). This point of view exploits in an essential way the so-called integration by parts formula which is used to derive a generalization of the classical Euler-Lagrange equation. As mentioned in V.3, this approach leads to conceptual difficulties.

A second approach, the so-called "Fluidodynamical picture", is based on a generalization of the classical Hamilton-Jacobi equation and inspired by the methods of stochastic control theory (see [62b] and also [90e]). Let us briefly present the basic facts of this method (more details can be found in the existing literature). To derive in classical mechanics the Hamilton-Jacobi equation from a variational principle, a velocity field $v(\cdot,\cdot)$ must be introduced such that $\mathring{q}(t) = v(q(t),t)$ and the equations of motion are

$$v(x,t) = \frac{1}{m} \nabla S(x,t) \tag{5.67}$$

$$\partial_t S + \frac{1}{2m} (\nabla S)^2 + V = 0. \tag{5.68}$$

As shown in Section I.3, the structure of the equations governing the Madelung fluid is not very different.

We consider stochastic diffusion processes $\{X_t\}_{t\in \mathbb{R}_+}$ with values in $\mathbb{R}^d$. We assume that the process starts at time $t = 0$ with a given density $\rho(\cdot,0)$ and evolves in time according to the stochastic differential equation

$$dX_t = b_+(X_t,t)dt + dW_t. \tag{5.69}$$

Like v in the classical framework, b_+ has to be now determined by dynamical constants. The critical diffusions make the stochastic action stationary with respect to the variation of the drift velocity field.

As explained in Chapter II we can also consider the backward velocity field $b_-(X_t,t)$ and we recall that

$$b_- = b_+ + \sigma^2 \frac{\nabla\rho}{\rho} \tag{5.70}$$

For the time interval [0,T] we can introduce the following action functional

$$A = \mathbb{E}\left[\int_0^T \left\{\frac{1}{2} m\, b_+(X_t,t)\cdot b_-(X_t,t) - V(X_t)\right\}dt\right] \tag{5.71}$$

where V is some scalar potential.

The variational principle gives conditions on the drifts b_+ and b_- based on the stationarity of the action functional, namely, $DA = 0$ under appropriate boundary conditions. One can show that critical diffusions have the following structure in terms of the density $\rho\,(\cdot,t)$ of the process and some auxiliary scalar function S

$$v(x,t) = \frac{1}{2}\,(b_+ + b_-) = \frac{1}{m}\,\nabla S \tag{5.72}$$

$$\partial_t(\rho v) = -\nabla(\rho v) \tag{5.73}$$

$$\partial_t S + \frac{1}{2m}\,(\nabla S)^2 + V - 2\nu^2 m^2\,\frac{\Delta\rho^{1/2}}{\rho^{1/2}} = 0\;. \tag{5.74}$$

This last equation is a stochastic generalization of the Hamilton-Jacobi equation, in which Bohm's quantum potential appears. In other words, one gets the Madelung fluid equations which, as discussed in Chapter I, are equivalent to the Schrödinger equation. The connection to quantum mechanics is expressed through the following formulae

$$2\nu = \frac{\hbar}{2m}\quad,\ \psi = \rho^{1/2}\,e^{\frac{i}{\hbar}S} \tag{5.75}$$

$$i\partial_t\psi = H\psi\;,\ H = -\frac{\hbar^2}{2m}\,\Delta + V\;. \tag{5.76}$$

As shown by E. Nelson [90e] one can also see that the mean action functional discussed above (and proposed in [62b]) is equivalent to a renormalized mean classical action integral.

An other approach to stochastic calculus of variations, the so-called "path-wise picture", is based on a generalization of the Euler-Lagrange equation. In this strategy we look at configurations making the mean classical action stationary with respect to the variation of the sample paths. To do this, a discretized version of the mean classical action is considered. Using then the elementary properties of the product of finite differences, one is able to calculate the variation of the mean classical action. For a complete account of this approach we refer to [60], [87a,b] (see also [81] for an application).

Let us emphasize an important difference between the two last approaches we briefly described. In the fluidodynamical picture the forward velocity field b_+ is constrained by construction to be a gradient, like the velocity field v in the corresponding classical situation. In the path-wise approach, the velocity fields are in general no longer the gradient of some scalar functions.

In classical mechanics the path is often obtained by using a variational method (Hamilton's principle) for a given pair of initial and final positions. J.C. Zambrini has given in a series of papers the constructive counterpart of these variational approach in Stochastic Mechanics [114b,c], realizing a program initiated in 1931 by E. Schrödinger [98b,c] by constructing a class of diffusion processes X_t indexed by $I = [-\frac{T}{2}, \frac{T}{2}]$ with two given probability densities $\rho_{-\frac{T}{2}}(x)$ and $\rho_{\frac{T}{2}}(y)$. The class of admissible processes in Zambrini's stochastic variational principle is not limited to Markovian processes, the Markov property being replaced by the Bernstein property, which is a one dimensional version of the local Markov property used for example in constructive field theory. Zambrini gives a new probabilistic interpretation of the heat equation which is the closest classical counterpart of quantum mechanics.

VI. TWO VIEWPOINTS CONCERNING QUANTUM AND STOCHASTIC MECHANICS

VI.1 General Remarks

There are three ways to look at stochastic mechanics.

a) Stochastic mechanics can be considered as a new quantization procedure leading to an alternative description of microphysics. Stochastic mechanics "attempts to provide a realistic objective description of quantum physical events in classical terms" (E. Nelson [90]), deriving the Schrödinger equation and all other elements of quantum physics from purely classical and probabilistic concepts. According to this viewpoint, the Schrödinger equation is not an equation of motion for the state of a physical system but a linear partial differential equation which enables one to determine the drift. The solution of the stochastic differential equation gives the trajectories of the diffusing particle. Nelson's background field hypothesis asserts that quantum fluctuations are the result of classical interaction with the background field (presumably electromagnetic). Indeed, no physical system of finitely many degrees of freedom can be considered as fully isolated and such a system is always in interaction with a background field.

b) The orthodox quantum mechanical description can be taken for granted and stochastic mechanics can be used to give a more detailed description of some aspects of microphysics. By "orthodox interpretation" we mean any interpretation which regards the wave function as a complete description of the state of a physical system. From this point of view, stochastic mechanics is a way of producing quantum mechanics, which assigns a Markov diffusion process in configuration space to each Schrödinger operator H and each quantum state ψ. Even if one does not accept stochastic mechanics as a reasonable physical theory, one cannot exclude the possibility that stochastic mechanics may provide new methods for studying quantum mechanical problems and that using the stochastic methods one arrives at new questions which are meaningless or unnatural for the standard interpretation of quantum mechanics. The first hitting time of a particle at a counter, a quantity which is not fully understood in quantum mechanics (see e.g. R. Werner [109]) becomes unambigously defined in the framework of stochastic mechanics (see e.g. A. Truman and J.T. Lewis [105]).

c) Stochastic mechanics can be viewed as a general description of a class of dynamical systems disturbed by some isotropic, translational invariant noice.

Lèt us finally remark that there exists a connection between the process associated to the Euclidean formulation of quantum mechanics obtained by analytical continuation to imaginary time and the process associated to the ground state by stochastic mechanics. The two processes are essentially the same since they are generated by unitarily equivalent semigroups. This intriguing observation was first noted by F. Guerra and P. Ruggiero [63]. See VI.9.

Depending on the attitude we choose to look at stochastic mechanics, the stochastic process appears as a representation of a physical reality or as a auxiliary mathematical tool which can be used to express physically relevant quantities and to prove theorems.

In this chapter, we will discuss briefly some topics in quantum theory from the point of view of stochastic mechanics. For greater details see [90a,b].

In the next chapter, we will consider other physical applications of stochastic mechanics which are not connected with quantum theory.

VI.2 Interference

If we send a beam of electrons through a screen with two slits in it, more generally a crystal, one observes a diffraction pattern. Its character can be predicted by quantum mechanics.

Let $\psi(x,t)$ the solution of the free Schrödinger equation on $\mathbb{R}^3$ for a particle of mass m

$$i\hbar\frac{\partial\psi}{\partial t} = -\frac{\hbar^2}{2m}\Delta\psi \tag{6.1}$$

with initial condition given by

$$\psi(x,0) = (2\pi\sigma^2)^{-3/4}\, e^{-\frac{x^2}{2\sigma^2}} \quad . \tag{6.2}$$

This solution corresponds to a Gaussian wave packet describing the diffraction by Gaussian slits of half width σ centered at the origin. To this solution the stochastic mechanics associates a Gaussian diffusion process which can easily be computed explicitly.

Introduce now the wave function

$$\phi(x,t) = N[\psi(x-a,t) + \psi(x+a,t)] \tag{6.3}$$

N being a normalization factor.

In a frame of reference moving with the beam, the function ϕ can be viewed as the wave function associated to the two-Gaussian slit experiment since the particle is at time $t = 0$ located at the part $\pm a$. Let $\lambda = \sigma m a^2/\hbar$ and $\alpha = ma/\hbar$, then the drift of the process associated to ϕ assesses the complicated form

$$b_+ = \frac{t-2\lambda^2}{4\lambda^4+t^2}\, x + \frac{\dfrac{t-2\lambda^2}{4\lambda^4+t^2}\sinh\dfrac{4\lambda^2\alpha\cdot x}{4\lambda^4+t^2} - \dfrac{t+2\lambda^2}{4\lambda^4+t^2}\sin\dfrac{2t\alpha\cdot x}{4\lambda^4+t^2}}{\cosh\dfrac{4\lambda^2\alpha\cdot x}{4\lambda^4+t^2} + \cos\dfrac{2t\alpha\cdot x}{4\lambda^4+t^2}}\, a \qquad (6.4)$$

and the corresponding process can no longer be computed explicitly. However, we can easily see that only the coordinate along the direction joining the two slits is relevant and that the drift is bounded, but the process comes very close to having nodes. Indeed, the following fact can be proved

a) If $t \ll \sigma \frac{ma^2}{\hbar}$ the process is pratically indistinguishable from the equally weighted mixture of the one slit processes starting from $+a$ and $-a$.

b) if $t \gg \sigma \frac{ma^2}{\hbar}$ the drift becomes very big near the lines $x = (2n+1)\frac{\pi\hbar}{2ma}\, t$, $n \in \mathbb{Z}$ and points away from them. It follows that the particle is confined in one of the regions between these lines and this trapping phenomenon produces the diffusion pattern.

VI.3 Observables - Measurement

VI.3a Observables

In quantum mechanics, to each observable corresponds a self-adjoint operator A. The quantum expectation value of the observable A in a state ψ is given by $\langle\psi, A\psi\rangle$.

In stochastic mechanics, observables correspond to the observation of the diffusion process X_t. The basic observations are given by a cylinder function of the form $f(X_{t_1}, \ldots, X_{t_n})$ f being real valued, bounded and measurable. In stochastic mechanics, an observable is a random variable. For position observables, at a fixed time, there is a one-to-one correspondence between the two descriptions. For the diffusion X_t associated with the wave function ψ, the expectation of X_t is $\langle\psi, X(t)\psi\rangle$, where $X(t)$ is the position operator in the Heisenberg representation. In other words, the Born interpretation is automatically verified in stochastic mechanics and is no longer a supplementary hypothesis, e.g. if ρ denotes the probability density of the process, one has $\rho(x,t)dx = |\psi(x,t)|^2 dx$.

For more general observables there exists no such one-to-one correspondence. Indeed, for a random variable which depends on more than one time the expectation of this random variable in general does not depend on a sesquilinear way on ψ and therefore no self-adjoint operator corresponds to such a random variable. For more details see [90e]. Moreover, the formalism of stochastic mechanics has been extended to the case of mixed quantum states and the relation between quantum mechanical and stochastic observables has been studied in this more general situation by R. Werner [109].

VI.3b Momentum Process

The privileged role of configurational variables in stochastic mechanics is in contrast to the L^2-formalism of quantum mechanics where one treats space and momentum coordinates on the same footing. F. Guerra and L. Morato [62a] have attempted to approach the position momentum complementarity in stochastic mechanics and managed to deal with the coherent states of the harmonic oscillator. By exploiting the full symmetry in position and momentum of the Hamiltonian, they could construct diffusion associated to momentum. Their strategy, however, does not seem to be able to generalize to other potentials.

Let us consider a point particle of mass m moving under the influence of a potential $V(x)$. In stochastic mechanics, several stochastic processes turn out to have mean values which coincide with the expectation of the quantum mechanical momentum operator P. For instance, we can take the forward and backward drifts on the current velocity

$$\mathbb{E}[b_+(X_t,t)] = \mathbb{E}[b_-(X_t,t)] = \mathbb{E}[v(X_t,t)] = \langle\psi(\cdot,t), P\psi(\cdot,t)\rangle . \quad (6.4)$$

But none of those random variables has the same distribution as the operator P, already their variance differs from those of P .

To introduce a momentum in stochastic mechanics, M. Davidson [30b], D. De Falco, S. De Martino and S. De Siena [35b] make use of the asymptotic behavior of the trajectories for a free particle.

Let X_t be the position process in a situation where a potential is present. Consider now the solution $\psi^{o,t}$ of the free Schrödinger equation with initial condition at time t being given by the interaction wave function ψ at time t:

$$\psi^{o,t}(x,t) = \psi(x,t) . \tag{6.5}$$

This leads to the free position process $X_T^{o,t}$ given by

$$dX_T^{o,t} = b_+^{o,t}(X_T^{o,t},T)\,dT + dW_T^{o,t} \tag{6.6}$$

where $W^{o,t}$ is a Wiener process with variance $\frac{\hbar}{m}$ (independent of $X_t^{o,t}$).

In particular, we can impose "per fiat" Davidson's construction [30b]

$$X_t^{o,t} = X_t \quad , \quad W_T^{o,t} = W_T . \tag{6.7}$$

The process $X^{o,t}$ can be thought of as being "tangent" to the process X at time t. Following [30b] and [35b] we define

$$\pi_t = \lim_{T\uparrow+\infty} m\,\frac{X_T^{o,t}}{T} . \tag{6.8}$$

According to the result discussed in Section VI.6, this limit exists a.s. and it has a probability density equal to the momentum distribution of the quantum state ψ. Thus in the case of arbitrary potential a random variable has been constructed whose distribution coincides with the momentum distribution in quantum mechanics.

In a recent paper, S. Golin [58b] has carefully analyzed this implementation of momentum in stochastic mechanics by discussing the ground state of the one dimensional harmonic oscillator

$$\psi(x,t) = (2\pi\sigma^2)^{-1/4}\exp\{\tfrac{1}{2}(i\omega t + \frac{x^2}{2\sigma^2})\} \tag{6.9}$$

where $\sigma^2 = \frac{\hbar}{2m\omega}$.

In this case, the momentum process π_t can be determined explicitly

$$\pi_t = m\omega e^{-\frac{\pi}{2}}[\xi_t + \int_0^t e^{\gamma(\tau-t)}dW_\tau] \tag{6.10}$$

where

$$\gamma(t) \doteq \arc \tan \omega t - \frac{1}{2}\log(1 + \omega^2 t^2)$$

and ξ_t is the position process solution of the stochastic differential equation

$$d\xi_t = -\omega\xi_t + dW_t \tag{6.11}$$

or, in integral form,

$$\xi_t = e^{-\omega t}[\xi_o + \int_0^t e^{\omega\tau}\, dW_\tau] \tag{6.12}$$

where W_t is the Wiener process with variance $\frac{\hbar}{m}$.

Golin's analysis pointed out some manifestly unphysical features of the momentum process. The most important ones are:

a) The momentum process π_t has no operational meaning, except in the free case. Indeed, whenever a non-vanishing potential is present, there is no experimental way of implementing the definition of π_t, because you cannot simply turn off the potential at time t. But this was required in the definition of π_t.

b) Using Itô's formula we can obtain a new representation of π_t

$$\pi_t = m\, e^{-\pi/2} \int_t^\infty [\omega - \dot{\gamma}(\tau-t)]e^{\gamma(\tau-t)}\, \xi_\tau\, d\tau \tag{6.13}$$

from which we can deduce that π_t has two continuous derivatives. Unfortunately, the derivative $\dot{\pi}_t$ of the momentum process cannot be interpretated as force. Indeed, the variance of $\dot{\pi}_t$ is different from the variance of the harmonic force.

c) There is no straight-forward way of giving the position-momentum incertainty relation using π_t (we shall discuss this point in Sec. VI.4).

From these unsatisfactory shortcomings of the process π_t Golin concludes that such a definition of momentum is unacceptable (see [58b]). More generally, unitary (canonical) transformations in quantum (classical) mechanics have to be replaced in stochastic mechanics by measure preserving transformation. A general discussion of covariance properties in the stochastic mechanical framework is still missing.

VI.3c Repeated Measurements: A Case Against Stochastic Mechanics?

As discussed in Section (VI.3a), for position measurements performed at a fixed time stochastic mechanics and quantum mechanics make exactly the same predictions. It was argued by H. Grabert, P. Hänggi and P. Talkner [58] and E. Nelson [90e,f] that the correlation for repeated measurements obtained in the framework of stochastic mechanics were in conflict with the quantum mechanical predictions. The aim of this Section is to prove that a careful consideration of the wave packet reduction in stochastic mechanics shows that in fact the quantum mechanical correlations can also be derived in the stochastic mechanical framework. We refer to [16] and [60b] for a more detailed discussion of measurement in stochastic mechanics.

Let us first sketch some apparent paradoxa appearing in stochastic mechanics in relation with the problem of repeated measurements.

a) Example 1. Consider two dynamically uncoupled harmonic oscillators O_1 and O_2 with circular frequency ω. We have two Hamiltonians H_1 and H_2 in the Hilbert spaces $\mathcal{H}_1 = \mathcal{H}_2 = L^2(\mathbb{R})$. The Hamilton operator of the total system is of the form

$$H = H_1 \otimes \mathbb{1}_2 + \mathbb{1}_1 \otimes H_2 \tag{6.14}$$

and acts on $\mathcal{H} = \mathcal{H}_1 \otimes \mathcal{H}_2$.

For any observable A_1 of $\mathcal{H}_1$ its time evolution in the Heisenberg picture is given by

$$e^{itH}(A_1 \otimes \mathbb{1}_2)e^{-itH} = e^{itH_1} A_1 e^{-itH_1} \otimes \mathbb{1}_2 \tag{6.15}$$

and is completely independent of the choice of H_2 since the systems are dynamically uncoupled. Let us perform a position measurement on O_1 at time $t = 0$ and a position measurement on O_2 at time $t > 0$. Since the corresponding Heisenberg position operators commute, $[X^1(0), X^2(t)] = 0$, the quantum mechanical correlation $\langle X^1(0)\, X^2(t)\rangle$ can be associated with this experiment. To carry out explicit computations, let us suppose that the state of the system is Gaussian. Then the stochastic mechanical correlation $\mathbb{E}[X_0^1\, X_t^2]$ is proportional to $e^{-\omega t}$ whereas the quantum mechanical correlation $\langle X^1(0)X^2(t)\rangle$ is periodic in t [16].

b) Example 2. A similar apparent paradox appears for a single harmonic oscillator in the ground state. Its stochastic mechanical correlation can easily be calculated

$$\mathbb{E}[X_0\, X_t] = \sigma^2\, e^{-\omega|t|}\ , \ \sigma^2 = \frac{\hbar}{2m}\ . \tag{6.16}$$

For $t = \frac{n\pi}{\omega}$, $n \in \mathbb{Z}$ the Heisenberg position operators $X(0)$ and $X(t)$ commute so that we may consider $\langle X(0)\ X(t)\rangle$ as the corresponding correlation. But

$$\langle X(0)\ X(t)\rangle = (-1)^n \sigma^2 \tag{6.17}$$

does not agree with the stochastic mechanical correlation which shows, as in Example 1, an exponential fall-off.

c) Example 3. Consider a particle in a scattering state and denote by P_{in} and P_{out} the initial and final momentum. The elasticity of the scattering is reflected in the fact that the corresponding kinetic energies commute $[P^2_{in}, P^2_{out}] = 0$ and quantum mechanical information about the scattering is contained in the correlation $\langle P^2_{in}\ P^2_{out}\rangle$. The corresponding random variables are defined by

$$\pi_{in} = \lim_{t\downarrow-\infty} \frac{mX_t}{t}, \quad \pi_{out} = \lim_{t\uparrow+\infty} \frac{mX_t}{t} . \tag{6.18}$$

As we will see in Section VI.6 the limits exist with probability one and have the correct quantum mechanical distributions. It turns out, however, that $\mathbb{E}[\pi^2_{in}\pi^2_{out}]$ is different from the quantum mechanical correlation.

Now, a resolution of these paradoxa is to be proposed. To be explicit, we consider only Example 2 since the others are resolved in a fully similar fashion. One of the basic features of stochastic mechanics is the dependence of the diffusion Markov processes on the drift

$$dX_t = b_+(X_t,t)dt + dW_t \tag{6.19}$$

where W_t is a Wiener process with variance νt. Thus, it seems natural in this framework, after a measurement on the system has been performed, to introduce a new process for the description of the system. We cannot, in fact, measure the correlation between the values of the process at different times because any attempt to localize the particle changes the drift. Therefore, after the first measurement at time $t=0$ we have automatically a new process. Suppose that the result of the first position measurement at time $t=0$ yields the value x_0. For $t > 0$, the new process $X_t^{x_0}$ is solution of the new stochastic differential equation

$$dX_t^{x_0} = b_+^{x_0}(X_t^{x_0},t)dt + dW_t^{x_0} \quad t > 0 \tag{6.20a}$$

$$\lim_{t \downarrow 0} X_t^{x_o} = x_o \quad \text{a.s.} \tag{6.20b}$$

where $W_t^{x_o}$ is a Wiener process with the same variance as W_t and with increments independent of those of W_t, $t \le 0$. The drift b^{x_o} is a functional of the quantum state. Denoting the quantum mechanical wave function after the measurement by $\phi_t^{x_o}$ with $\lim_{t \downarrow 0} \phi_t^{x_o} = \delta(x-x_o)$, then we have

$$b_+^{x_o} = \frac{\hbar}{m} (\mathrm{Re} + \mathrm{Im}) \nabla \log \phi_t^{x_o} . \tag{6.21}$$

The probabilistic information about repeated position measurement at time $t > 0$ is entirely contained in $X_t^{x_o}$, whereas X_t is in this context of no significance whatsoever. In this way, the wave packet reduction has naturally been incorporated into stochastic mechanics. According to this analysis it is obviously not the auto-correlation function $\mathbb{E}[X_o X_t]$ but the quantity

$$\int dX_o \, \rho(x_o, 0) \, x_o \, \mathbb{E}[X_t^{x_o}] \tag{6.22}$$

that gives the prediction for the correlation experiment. Indeed, we now get agreement with the quantum mechanical correlation.

The solution $\phi_t^{x_o}$ of the Schrödinger equation (3.24) for the harmonic oscillator takes the form

$$\phi_t^{x_o}(x) = \int dx' \, K_t(x,x') \, \phi_o^{x_o}(x') = K_t(x,x_o) \tag{6.23}$$

where the kernel $K_t(x,x')$ is given by the following explicit formula

$$K_t(x,x') = \left(\frac{m\omega}{i2\pi\hbar \sin\omega t} \right)^{1/2} \exp\left\{ - \frac{m\omega}{2\hbar} (x^2 - x'^2) - \frac{m\omega}{\hbar} \frac{(e^{-i\omega t} x - x')^2}{1 - e^{-2i\omega t}} \right\} . \tag{6.24}$$

Hence the drift $b_+^{x_o}$ takes the form

$$b_+^{x_o}(x,t) = \frac{x}{\tan\omega t} - \frac{x_o}{\sin\omega t} . \tag{6.25}$$

Consequently, the stochastic differential (6.20) is linear and can be solved

$$X_t^{x_o}(s) = (\cos\omega t - \sin\omega t \, \mathrm{cotg}\,\omega s) \; x_o + \frac{\sin\omega t}{\sin\omega s} X_s^{x_o} + \sin\omega t \int_s^t \frac{dW_\tau^{x_o}}{\sin\omega\tau}$$

$$s \le t, \quad \sin\omega s \ne 0 . \tag{6.26}$$

For t being a multiple of $\frac{\pi}{\omega}$, the random variable $X_t^{x_0}$ is just the constant $(-1)^n x_0$ a.s. Thus the correlation (6.22) is simply

$$(-1)^n \int dx_o \, \rho(x_o,0) x_o^2 = (-1)^n \, \mathbb{E}(X_0^2) \tag{6.27}$$

and it coincides with the quantum mechanical correlation.

VI.4 Indeterminacy Relations

The earliest version of an indeterminacy relation in quantum mechanics is due to W. Heisenberg [64] in 1927.

Let A and B be two Hermitian operators such that

$$[A,B] = c \, \mathbb{1} \tag{6.28}$$

where $c \in \mathbb{C}$, then Heisenberg proved that

$$\mathrm{Var}\, A \cdot \mathrm{Var}\, B \geq \frac{1}{4} c^2 \tag{6.29}$$

where

$$\mathrm{Var}\, A \equiv \langle A - \langle A \rangle \rangle^2 \tag{6.30}$$

and $\langle \cdot \rangle$ denotes as usual the quantum mechanical expectation.

In 1930, E. Schrödinger [98a] established a stronger form of (6.29). Defining the covariance of the operators A and B by

$$\mathrm{Cov}(A,B) \equiv \frac{1}{2} \langle AB + BA \rangle - \langle A \rangle \langle B \rangle \tag{6.31}$$

then Schrödinger's version of indeterminacy relation takes the form

$$\mathrm{Var}\, A \cdot \mathrm{Var}\, B \geq \mathrm{Cov}^2(A,B) + \frac{1}{4} |\langle [A,B] \rangle|^2 \tag{6.32}$$

which is clearly stronger than (6.30).

Several indeterminacy relations can be derived in the stochastic framework. As discussed in Chapter I, their existence is a characteristic feature of diffusion processes. In 1930, R. Fürth [51b] derived a position velocity uncertainty relation for the heat equation, i.e. for the Wiener process. I. Fenyes [45], L. de la Pena Auerbach and M. Cetto [36], D. de Falco, S. de Martino and S. de Siena [35] have obtained stochastic mechanical indeterminacy relations. As proved recently by S. Golin [58a,c] the indeterminacy relations which can be derived in stochastic mechanics are fully equivalent to Schrödinger's version of quantum mechanical indeterminacy relations.

We consider the case where the diffusion coefficient is a constant ν. Let f be a function of space and time. Denoting as usual by u the osmotic velocity, the following formula is obtained by integration by part

$$\mathbb{E}[f\,u] = -\nu\,\mathbb{E}[\nabla f]\ . \tag{6.33}$$

Using now the Schwarz inequality and the fact that the osmotic velocity u has zero mean, we obtain

$$\text{Var } f \quad \text{Var } u = \mathbb{E}[(f-E(f))^2]E[u^2]$$
$$\geq \mathbb{E}^2[(f-E(f))u] = E^2[f\,u]$$

then

$$\text{Var } f \text{ Var } u \geq \nu^2\ \mathbb{E}^2[\nabla f]\ . \tag{6.34}$$

If we set $f(x) = x$, we obtain now

$$\text{Var } x \text{ Var } u \geq \nu^2 \tag{6.35}$$

and by means of the Schwarz inequality (v being the current velocity)

$$\text{Var } x \text{ Var } v \geq \text{Cov}^2(x,v)\ . \tag{6.36}$$

This now yields the following position - momentum indeterminacy relation in stochastic mechanics

$$\text{Var } x(\text{Var } u + \text{Var } v) \geq \text{Cov}^2(x,v) + \nu^2\ . \tag{6.37}$$

The distribution of the diffusion process X_t and of the quantum quantum mechanical operator X_{op} coincides. Moreover, the momentum operator P_{op} satisfies

$$\text{Var } P_{op} = m^2(\text{Var } u + \text{Var } v) \tag{6.38}$$

$$\text{Cov}(X_{op},P_{op}) = m\ \text{Cov}(x,v)\ . \tag{6.39}$$

Therefore, the above stochastic mechanical indeterminacy relation is equivalent to Schrödinger's stronger version of the position-momentum uncertainty relation in quantum mechanics

$$\mathrm{Var}\, X_{op}\ \mathrm{Var}\, P_{op} \geq \mathrm{Cov}(X_{op}, P_{op}) + \frac{\hbar^2}{4} \tag{6.40}$$

by setting the diffusion constant $\nu = \frac{\hbar}{2m}$.

In [68a] and [58c] force-momentum uncertainty relation, angle variables - orbital angular momentum indeterminacy relations, time-energy indeterminacy relations are discussed in the framework of stochastic mechanics. It is worthwhile mentioning that all these uncertainty relations are a general feature of stochastic systems (diffusions) and that the diffusion constant ν could be any positive constant. In fact, the indeterminacy relations depend on a purely kinematical feature of diffusions, namely the non-differentiability of their sample path.

Remark 1. Reversing the point of view, it is natural to ask the following question: Given the quantum mechanical uncertainty relation, what can we infer about the notion of the quantum particle? As discussed by L.F. Abott and M.R. Wise [1] the Heisenberg position-momentum uncertainty principle is reflected in the fractal nature of the quantum mechanical paths, viz. the paths have Hausdorff dimension 2. The Hausdorff dimension of a closed set A in $\mathbb{R}^d$ can be defined in the following way. Let $h(t)$ be an increasing function of $t > 0$ either concave or convex. The Hausdorff measure of A with respect to h is strictly positive (may be infinite) if and only if A carries a positive measure $\mu \neq 0$ such that for all balls B of diameter $|B|$, $\mu(B) \leq h(|B|)$. The Hausdorff dimension of A is the supremum of the $\alpha \geq 0$ such that $h(t) = t^{\alpha}$ has this property. But this is exactly the regularity property of the sample paths of diffusion processes. The Wiener process has Hölder continuous paths of any order $\alpha < 1/2$ (see Chapter II).

Remark 2. We can ask for the position-momentum uncertainty relation using π_t as defined in Section (VI.3b). Using Schwarz inequality, we get

$$\mathrm{Var}\, x\ \mathrm{Var}\pi \geq \mathrm{Cov}^2(x,\pi) \ . \tag{6.41}$$

For the ground state of the harmonic oscillator

$$\mathrm{Cov}^2(x,\pi) = \frac{\hbar^2}{u}\ e^{-\pi} \tag{6.42}$$

which does not coincide with $Cov(X_{op}, P_{op}) = 0$ in this case.

VI.5 Locality

The causality principle asserts that any physically real phenomenon cannot be affected by a disturbance which occurs later in time. If we accept relativity theory, this implies the locality principle: Any physically real property cannot be influenced by something that occurs outside its backward light cone.

The experimental results of quantum mechanics are subject to randomness and there are correlations in the results of measurements on widely separated particles which have interacted in the past (see e.g. [43]).

Bell's inequality [11] is the most dramatical illustration of the relation between probability theory and quantum mechanics. This inequality is a constraint which has to be satisfied by any purely probabilistic model of discrete spin. This inequality is violated in quantum mechanics, which implies that quantum mechanics has no underlying probability model

The locality principle and the experimental confirmation of the prediction of quantum mechanics (and also of stochastic mechanics) forces us to conclude that determinism is ruled out and that there is an intrinsic randomness in nature which has nothing to do with our ignorance of the initial data.

Independently of the nature of space time, locality can be discussed in terms of separability of correlated but dynamically uncoupled systems. In quantum mechanics, if there is no quantum mechanical interaction between two systems and if we are only interested in observables of system 1, we may ignore system 2 completely as explained in Section (VI.3c). This very convenient feature of quantum mechanics is no more satisfied in stochastic mechanics. Nelson [90c] gives an example of a system for which the autocorrelation $\mathbb{E}[X_t^1 \; X_t^2]$ depends on the choice of the Hamiltonian H_2 of the second system. This is due to the fact that the diffusion takes place on configuration space $M = M_1 \times M_2$ and that both components of the drift depend, in general, on the total configuration. The stochastic mechanics is non-local: even if the particles are arbitrarily far separated, the first one is affected by the second.

VI.6 Scattering Theory

In scattering experiments, the asymptotic momentum is not measured directly, one is only able to measure positions and times. To determine the final momentum, one can use the follwoing method. If the particle was close to the scattering center at time 0 and if it is detected in a counter at the point $x \in \mathbb{R}^3$ at time $T > 0$ and if the distance between the scattering center and the place of detection is much greater than the range of interaction, it is reasonable to assume that during most of its flight the scattered particle moved nearly freely with a momentum close to P_f. This implies that $P_f \sim \frac{x}{T}$, for a particle of mass 1. Therefore, in stochastic mechanics it is natural to study the time evolution of $\pi_t = \frac{1}{t} X_t$. Given a potential $V(x)$ we have to consider that diffusion which can leave the region where the potential is strong and to show for such diffusion that for process π_t the following limit

$$\lim_{t\uparrow+\infty} \pi_t(\omega) = \lim_{t\uparrow+\infty} \frac{1}{t} X_t(\omega) = P_f(\omega) \tag{6.43}$$

exists pathwise with probability one. D.S. Schuker [99] has proved such a result in the free case where $V \equiv 0$ and E. Carlen [22c,e] has proved for a large class of potentials (potentials of Kato-Rellich type) that the random variable P_f exists almost surely, is square integrable and has the same distribution as the quantum mechanical final momentum for the corresponding solution ψ of the Schrödinger equation. For large t, π_t is a measure of the momentum and since $|\psi(x,t)|^2$ is the probability density of X_t, so is $|\psi(pt,t)|^2 t^3$ the probability density of π_t. A simple calculation shows that

$$\lim_{t\uparrow+\infty} t^3|\psi(pt,t)|^2 = |\tilde{\psi}(p)|^2 \tag{6.44}$$

where $\tilde{\psi}$ is the Fourier transform of ψ.

Let us now consider a diffusion process in $\mathbb{R}^3$ such that

$$\lim_{t\uparrow+\infty} \frac{1}{t} X_t(\omega) \equiv v_+(\omega) \tag{6.45}$$

exists almost surely. Let $\mathcal{B}_+ = \bigcap_{t>1} \sigma\{X_u | u > t\}$ be the tail field associated to this process. Clearly, v_+ is $\mathcal{B}_+$-measurable. It is natural to ask the following question: Can v_+ generate the whole tail field $\mathcal{B}_+$? If this is the case then any bounded $\mathcal{B}_+$-measurable random variable F admits the representation

$$F(\omega) = f(v_+(\omega)) \quad \text{a.s.}$$

for some bounded Borel function f on $\mathbb{R}^3$. This question is physically very important. Indeed, if v_+ does not generate the tail field $\mathcal{B}_+$, this implies that there exists extra scattering information besides the final momentum which can be gained by observing only the large time behaviour of the sample paths of the diffusion process. In a very interesting paper E. Carlen [22c] - using coparabolic Martin representation methods - proved that for a large class of potentials V the tail field $\mathcal{B}_+$ associated to the diffusion process of stochastic mechanics is indeed generated by v_+. In stochastic mechanics the scattering observables correspond to $\mathcal{B}_+$-measurable functions and in quantum mechanics the only scattering observables are functions of the quantum mechanical momentum operator $\frac{\hbar}{i}\nabla$ which is the operator theoretic analogon of the statement that the tail field is generated by v_+. By Carlen's result $\mathcal{B}_+$ does not contain any extra information, which agrees with the answer given by quantum mechanics.

Using other methods, let us mention the results of P. Biler [12] in the one dimensional case and M. Serva [96] for central potentials which both discuss potential scattering in stochastic mechanics.

Nelson [90e] considered a Gaussian wave packet under the free evolution and computed the correlation matrix of the initial momentum and the final momentum and found it to be $-e^{-\pi}\mathbb{1}$, independent of the width of the Gaussian. Therefore, the two momenta differ although their density coincides and this result shows the difficulty of defining a pathwise analogon of the scattering matrix in stochastic mechanics. Similarly, the correlation coefficient of the square of the momenta is equal to $-e^{-2\pi}$. Hence, there is no pathwise energy conservation, i.e. the trajectories of the configuration process do not exhibit elastic scattering.

Let us mention finally that there is a case in which a direct relation between scattering quantities and probabilistic quantities come out naturally, namely the limit of low energies in which the quantum mechanical cross section is given by a geometrical quantity, the scattering length. In [5] the relation between the asymptotic behaviour for $|x| \to +\infty$ of the drift associated by the Dirichlet form approach to quantum mechanics with the Schrödinger operator $H = -\Delta + V$ through $b_+ = \nabla \log \varphi_0^2$, $H\varphi_0 = E_0\varphi_0$, φ_0 being the ground state wave and the spectral properties of H at E_0 is discussed. It is also shown that the leading term in the asymptotic behaviour of b_+ for

$|x| \to +\infty$ determines the scattering length a but gives no information about the effective range parameter of the potential V. For more details, see [5]. To obtain complete information about the scattering, the processes associated to the excited states must be considered.

VI.7 Spinning Particle

One attempt to describe particles with spin in stochastic mechanics is based on the Bopp-Hagg model [29b] which interprets spinning particles as quantum rigid bodies. In this framework, the configuration space of a point particle with orientation is the manifold $M = \mathbb{R}^3 \times SO(3)$, which has the universal covering space $\tilde{M} = \mathbb{R}^3 \times SU(2)$, $\tilde{M}$ being a double covering of M. Our goal is to construct diffusion Therefore the following question must be answered: what are the smooth wave functions ψ on M which give rise to a diffusion on M? It turns out that ψ must be either an integral spin wave function or a half-integral spin wave function. A superposition of wave functions of the two classes does not correspond to a diffusion on M.

The theory of spin in stochastic mechanics has been elaborated by T.G. Dankel [29b] and D. Dohrn, F. Guerra, P. Ruggiero [40].

In absence of electromagnetic field the Schrödinger equation will be

$$i \hbar \frac{\partial \psi}{\partial t} = -\frac{\hbar^2}{2} \Delta \psi \tag{6.46}$$

where Δ is the Laplace-Beltrami operator on $\tilde{M}$ involving constants related to the mass m and the moment of inertia I of the particle, we assume spherical symmetry for the sake of simplicity. In a non-relativistic theory, the moment of inertia of a point particle is 0. To take this fact into account, the limit $I \to 0$ must be taken. In classical mechanics this limit exists, but it is uninteresting since the translational degrees of freedom and the rotational ones fully decouple, which express the well-known fact that there is no classical deterministic analogon of spin. Dankel shows that if ψ transforms according to a spin s representation of $SU(2)$, the quantum mechanical limit $I \to 0$ also exists and the corresponding wave function ψ verifies the Pauli equation for spin s with multiplicity $2s+1$. The question of the existence of the limit $I \to 0$ in stochastic mechanics is open. For a heuristic argument, see [90e]. Let us also mention that W. Faris [44] has shown that stochastic mechanics gives a perfectly consistent probabilistic description of the Einstein-Rosen-Podolsky-Bohm experiment, a more practical version of E.P.R. experiment that involves spin.

Let us now briefly mention a more pragmatic point of view to introduce spin in a stochastic framework. For a spin 1/2-particle, this approach starts from quantum mechanics and tries to interpret the continuity equation for $|\psi_t(x,\sigma)|^2$, $\sigma = \pm 1$, as a forward Kolmogorov equation. In this procedure, to each smooth solution without nodes is associated a Markov process

$$Y_t = \{(X_t,\sigma_t) \in \mathbb{R}^3 \times \{-1,1\}\}$$

which reproduces quantum averages for coordinates and a selected component of the spin which is treated as a discrete random variable. For more details, we refer to [31], [32], [33].

This general heuristic principle which is based on the identification of the quantum mechanical continuity equation for $\rho_t = |\psi_t|^2$ with forward Kolmogorov equations for suitably chosen random processes, is also useful in other physically interesting cases (see, e.g. [14b,c]) [26], [32]).

VI.8 Pauli Principle

Let us consider in $\mathbb{R}^3$, N particles which cannot be distinguished. The configuration space of this system is the manifold M consisting of all unordered N-uples $\{X_1,\ldots X_N\}$ in $\mathbb{R}^{3N}$, where the X_i are distinct points of $\mathbb{R}^3$. This is a differentiable manifold which is not simply connected if $N > 1$. Indeed, the universal covering space of M is $\tilde{M} = \mathbb{R}^{3N}/D$, D being the set of all ordered N-uples such that two or more points coincide. The fundamental group of M is the symmetric group S_N. To construct diffusion on M we ask when a smooth wave function ψ on $\tilde{M}$ generates a diffusion on M. This is the case if the wave function ψ is either symmetric or antisymmetric but not a superposition of the two. It follows from this result that the exclusion principle, e.g. the distinction between Bose-Einstein and Fermi-Dirac statistics is a consequence of the basic principle of stochastic mechanics and is not an additional hypothesis. For more details, we refer to [90e], [66]

VI.9 The Connection Between Stochastic Mechanics and Euclidean Quantum Mechanics

The Euclidean formulation of quantum mechanics is obtained by analytical continuation to imaginary time. The existence of such analytical continuation follows from the positivity of the Hamiltonian. The time evolution is now given by the semi-group e^{-tH}. In this framework, the Schrödinger equation is replaced by a diffusion equation and therefore a stochastic interpretation is very natural and suggestive. However, it should be emphasized that in this approach the diffusion processes play a purely auxiliary role since they do not take place in "real time" and are only used as a mathematical tool to prove theorems about operators on Hilbert space (see e.g. [57], [100a].

Let ψ_0 be the ground state of the Hamiltonian

$$H = -\frac{\hbar^2}{2m}\Delta + V \tag{6.47}$$

$$H\psi_0 = 0 . \tag{6.48}$$

Since ψ_0 is strictly positive it follows that the current velocity $v = 0$ and the associated process which is solution of the stochastic differential equation [60a], [95]

$$dX_t = b_+(X_t)dt + \frac{\hbar}{2m}dW_t$$

is stationary.

Let us consider as simplest example the ground state of the one-dimensional harmonic oscillator

$$\psi_0 = (2\pi\sigma)^{-1/4} e^{-\frac{x^2}{4\sigma}} \qquad \sigma = \frac{\hbar}{2m\omega} \tag{6.49}$$

which leads to the drift vector

$$b_+ = -\omega x \tag{6.50}$$

and to the Fokker-Planck equation

$$\frac{\partial\rho}{\partial t} = \frac{\hbar}{2m}\frac{\partial^2\rho}{\partial x^2} + \omega\rho + \omega x\frac{\partial\rho}{\partial t} . \tag{6.51}$$

This equation can explicitly be solved. Namely, we have

$$\rho(x,t) = \int_{\mathbb{R}} \rho(y,0) \quad p(y,0,x,t)dy \tag{6.52}$$

with

$$p(y,0,x,t) = [2\pi\bar{\sigma}(t)]^{-1/2}\exp\{-\frac{1}{2\pi\bar{\sigma}(t)}(x-e^{-\omega t}y)^2\} \quad . \tag{6.53}$$

An easy calculation shows that

$$\mathbb{E}[X_o X_t] = \bar{\sigma}(t) = \sigma e^{-\omega|t|} \tag{6.54}$$

which looks very "Euclidean".

More generally, for the ground state process the Fokker-Planck equation can be written

$$\frac{\partial\rho}{\partial t} = \frac{\hbar}{2m}\Delta\rho - \frac{\hbar}{2m}\operatorname{div}\left[(\frac{\nabla\rho_o}{\rho_o})\rho\right] \tag{6.55}$$

and has the stationary solution $\rho = \rho_o = \psi_o^2$.

It turns out that in this case

$$\mathbb{E}[X_o X_t] = \langle\psi_o, X e^{-|t|\frac{H}{\hbar}} X \psi_o\rangle_{L^2(\mathbb{R}^d)} \quad . \tag{6.56}$$

This relation is very remarkable since it links a "real time" quantity on the left hand side to an "imaginary time" quantity on the right hand side. This connection was first noted by F. Guerra and P. Ruggiero [63]: "Euclidean quantum mechanics (or field theory) is the ground state process of stochastic mechanics".

To show the mathematical origin of this relation, let us remark that the Fokker-Planck operator H_{FP} given by

$$H_{FP} = -\frac{\hbar^2}{2m}\Delta + \hbar\nabla.(b_+.) \tag{6.57}$$

with

$$b_+ = \frac{\hbar}{2m}\frac{\nabla\psi_o}{\psi_o} \tag{6.58}$$

is not only symmetric in $L^2(\mathbb{R}^d,\rho_o dx)$ but actually unitarily equivalent to the self-adjoint Hamiltonian operator H (defined in $L^2(\mathbb{R}^d,dx)$)

$$H_{FP} = U_{\psi_o}^{-1} H U_{\psi_o} \tag{6.59}$$

where U_{ψ_0} is the unitary operator from $L^2(\mathbb{R}^d, \rho_0 dx)$ onto $L^2(\mathbb{R}^d, dx)$ given by multiplication by ψ_0.

From this it follows that the Fokker-Planck equation has the formal solution

$$\begin{aligned}\rho(x,t) &= e^{-\frac{t}{\hbar} H_{FP}} \rho(x,0) \qquad\qquad t > 0 \\ &= U_{\psi_0}^{-1} \; e^{-\frac{tH}{\hbar}} U_{\psi_0} \; \rho(x,0) \\ &= \psi_0 \; e^{-\frac{tH}{\hbar}} \psi_0^{-1} \; \rho(x,0) \; .\end{aligned}$$

Moreover, this relation shows that the Fokker-Planck equation describes a relaxation process, the relaxation times being directly related to the spectrum of H. The unitary equivalence of H and H_{FP} tell us that the study of the unitary operator $e^{-itH/\hbar}$ in quantum mechanics can naturally be connected with the study of the Markov semigroup $e^{-tH_{FP}/\hbar}$(homomorphically extended to $e^{-itH_{FP}/\hbar}$, unitarily equivalent to $e^{-itH/\hbar}$). The associated process is called a distorted Brownian motion (see [8]).

<u>Remark</u>: The relation linking the real time quantity $\mathbb{E}[X_0 X_t]$ with the imaginary time object $\langle \psi_0, x e^{-tH/\hbar} x \psi_0 \rangle_{L^2(\mathbb{R}^d)}$ can be true only since the real time quantity is not accessible to measurement.

VI.10 The Semiclassical Limit

An approach based on stochastic mechanics has been very useful to study certain aspects of the semiclassical limit of quantum mechanics, i.e. the limit $\hbar \to 0$. In this limit the stochastic differential equation

$$dX_t = b_+(X_t,t)dt + \left(\frac{\hbar}{m}\right)^{1/2} dW_t \tag{6.61}$$

can be analyzed using the theory of large deviations for stochastic processes [49],[103],[107]. Adapting the Freidlin and Ventzel method. G. Jona-Lasinio, F. Martinelli and E. Scoppola [79] have discovered new, very interesting features of the semiclassical limit in the stationary case like the tunneling instability due to localized deformation of a multiwell potential exhibiting several equal minima (i.e. the classical ground state is degenerate).

The method consists in studying the process associated to the ground state ψ_0 of the quantum system. In this case, b_+ is a gradient

$$b_+ = \frac{\hbar}{2m} \nabla \log \psi_0^2 \tag{6.62}$$

and from the equation (2.52) we conclude that

$$\frac{\hbar}{m} \nabla \cdot b_+ + b_+^2 = \frac{2(V-E)}{m} \tag{6.63}$$

and can separate the problem in two steps. The first one consists in studying the solution of equation (6.63) when $\hbar \to 0$. Indeed, the logarithmic derivative of the ground state wave function contains the essential information on the tunneling. The second step consists in computing by probabilistic methods the spectrum of the generator A of the process which is associated to the Hamiltonian by $-\hbar A + E_0 = \psi_0^{-1} H \psi_0 = H^*_{FP}$ by (6.59), in the limit $\hbar \to 0$, to estimate the splitting of the ground state level. The results obtained show quite generally that both the localization state of the wave function and the splitting of the ground state are very sensitive to small local deformations of the potential. In particular, this is the position of the deformation rather than its absolute value which is the relevant factor. This work was extended recently in various directions [72], [100b].

More generally, in the framework of stochastic mechanics the semiclassical limit consists in comparing a classical smooth path with a diffusing one in its neighbourhood.

In the weak noise limit (for example as $\hbar \to 0$) the theory of large deviations [103] [107] leads in the simplest case a behaviour of the form $e^{-S/\hbar}$ which can not be handled by usual perturbation theory, indicating how it is natural to use the methods of stochastic mechanics to study non-perturbative effects.

VI.11 Bose Quantum Field Theory

Guerra and Ruggiero [60 d] [63] have investigated quantum fields from the point of view of stochastic mechanics. Enclosing the free scalar field into a large but finite box $B \subset \mathbb{R}^3$ and expanding it using a complete orthonormal basis the study of the free field is reduced to the study of an assembly of independent harmonic oscillators, each of which performs the diffusion associated with the ground state wave function.

Removing the cut-off i.e. in the limit $B \to \mathbb{R}^3$ Guerra and Ruggiero found that the ground state process for a scalar free field is the free Euclidean Markov field with mean zero and covariance $S(x-y)$ given by

$$S(x-y) = E\,[(\varphi(x,t)\;\varphi(y,t)]$$
$$= \frac{1}{(2\pi)^3} \int_{\mathbb{R}^3} e^{ik.(x-y)} \frac{dk}{2\omega(k)}$$

with $\omega(k) = (k^2 + m^2)^{1/2}$.

The underlying stochastic differential equation for the field φ of mass m can be written

$$d\varphi(x,t) = -(-\Delta_x + m^2)^{1/2} \varphi(x,t)\,dt + dW(x,t)$$

where $W(x,t)$ is such that

$$E\,[\,dW(x,t)\;dW(y,t)\,] = \delta(x-y)\,dt$$

This gives a new interpretation of the free Euclidean Markov field in real time. For more details see [63].

In [90 f] Nelson extends the Guerra-Ruggiero procedure to construct a family of random fields φ_j for scalar currents j with the linear coupling $j(x)\;\varphi(x)$. Moreover he suggests that no real understanding of quantum mechanics is possible without considering the larger framework of field theory.

In a recent paper [22 f] E. Carlen uses the Schrödinger representation of the quantum dynamics for free fields to construct the stochastic mechanics of the free scalar field of mass m and to investigate the sample path properties of the stochastic mechanical diffusions corresponding to single particle configurations of the field.

In this framework the Klein-Gordon equation

$$\left(\frac{\partial^2}{\partial t^2} - \Delta_d + m^2\right) \varphi = 0$$

is a kinematical equation which stands in the same relation to Schrödinger equation for the quantized field as does the Newton equation for a free particle in $\mathbb{R}^d$ $m\ddot{x} = o$ to the free Schrödinger equation for a particle of mass m in $\mathbb{R}^d$. In both cases the kinematical equation is used to define the classical phase space of the system.

Very recently Ph. Blanchard, E. Carlen and G.F. Dell'Antonio have shown that one particle states in which the corresponding solutions of the Klein-Gordon equation are strongly localized in the Newton-Wigner sense have as their filtered field configurations (obtained by filtering out the vacuum fluctuations in a physically meaningful covariant way) functions with one localized bump localized near the Newton-Wigner position. A detailed account of the considerations sketched here can be found in [13 bis].

VII. A NON-QUANTAL LOOK AT STOCHASTIC MECHANICS

VII.1 General Remarks

This chapter is devoted to the possibility of extending Nelson's theory, which originally is a derivation of quantum mechanics from the classical one, to provide a general mathematical framework adapted to the description of a class of dynamical systems randomly disturbed. In classical physics there is a large class of problems, which are well modelized in terms of stochastic processes. Let us consider a system with many degrees of freedoms, such that it is possible to select a small number of degrees of freedoms the variations of which are slower than for the other. Incorporating now all the fast degrees of freedom in a noisy source we have only to consider the equations of motion of the slower one. These equations of motion become then stochastic differential equations. In a phenomenological description of this type the fluctuations are the result of the interaction with the enormous number of degree of freedom of the environment.

Let us now be more precise about the general physical basis of our model. We consider a large number of particles travelling in a conservative free field. The motion of an individual particle is quite well understood and the classical deterministic equations of motions are given by

$$x_i = \frac{1}{m} p_i$$

$$p_i = -\nabla + F(x_1, \ldots, x_N) \qquad (7.1)$$

F describing the interaction with the other particles. The large number of particles involved justifies a statistical treatment. But furthermore if on large scale the system is stable, on much smaller scale this is not the case, local irregularities of the force field as well as collisions and nearby encounters of particles tend to modify in a random way the classical picture described before. They are the source for a random behaviour of the particles.

Both these reasons justify a statistical model in which the trajectories of the particles are modeled by the paths $x_t(\omega)$ of a stochastic process. In many situations from a physical point of view it seems natural to assume that the stochastic process gives zero weight to discontinuous paths. Furthermore it seems convenient and reasonable to consider only Markov random processes i.e. those processes for which the frequence of random changes due to local irregularities and collisions makes that the particle keeps only the memory of the past through its present state. These rather innocent assumptions imply mathematically that the random process we consider is a diffusion process. In other words the process is solution of the stochastic differential equation

$$dX_t = b_+(X_t,t)\ dt + \sigma\ dW_t \tag{7.2}$$

where X_t is its position at time t, b_+ a velocity field and W_t the standard Wiener process. σ is a diffusion coefficient taking into account the diffusive properties of the environment. Strictly speaking (7.2) implies that we made the additional assumption of homogeneous, isotropic and constant diffusion (see section II.3). The model we have in mind can be generalized to take into account more general situations. However at this stage is is not necessary to consider such a refinement.

Up to now the drift field b_+ is unspecified. Indeed, if the short range forces have been taken into account by the diffusion coefficient σ , the influence of the deterministic force field $-\nabla V$ has not been considered. The paths of the process are not differentiable, a fact which reflects the random character of the environment on a small scale. However, if one reminds the smooth character of the force field on a much larger scale, one can define a stochastic acceleration obtained by a generalization of the notion of derivative for a diffusion process (see Chapter II), which allows to write a Newton's law in the mean:

$$ma = -\nabla V(x) \tag{7.3}$$

According to this procedure the underlying stochastic equations are constructed

a) by generalizing the classical kinematics in order to allow for probabilistic description

b) by a generalization of the classical dynamical law appropriate for diffusion motion, which gives to the drift b_+ a dynamical meaning. In other words the stochastic Newton law is a bridge between the disorder existing at the microscopic scale and the overall force field acting at large scale.

The next step consists in investigating the properties of the probability density of the processes i.e. the functions $\rho(x,t)$, $x \in \mathbb{R}^d$ such that

$$\mathbb{E}\,[f(X_t)] = \int_{\mathbb{R}^d} \rho(t,x)\ f(x)dx \tag{7.4}$$

where $\mathbb{E}$ denotes the expectation with respect to the random process X_t. The density ρ satisfies the Fokker-Planck equation and an additional constraint coming from the Newton's law in the mean. To solve explicitly this couple of non-linear equations involving ρ and b_+ it is convenient to suppose that the current velocity (the part of b_+ reversing its sign under time reversal) is a gradient field (see Chapter III). In a sense this further assumption corresponds to situations where noisy turbulence is on a much smaller scale than the features we want to describe. The current velocity is expected to be observable. It must be emphasized at this point that the equations we obtain, although identical in form to the equations of Nelson's stochastic mechanics, have a different physical meaning. In Nelson's approach no statement is made about the physical nature of the noise, neither is needed. In our model the mechanism responsible for the diffusion arises from a real physical process.

Let us also remark that the conventional description of such systems by diffusions in velocity space is not contradictory to our proposal. Indeed, appealing to the formal analogy of our model with quantum mechanics one can realize that if the diffusion constant, which plays the rôle of the Planck constant $\hbar$, is very small the stochastic process X_t depresses in a exponential way by a factor of the form $e^{-\frac{S}{\sigma}}$ the weight of those paths which are far from the "classical" ones, i.e. those corresponding to $\sigma = 0$, which are solutions of (7.1). Indeed, random trajectories become deterministic in the limit $\sigma \to 0$.

Also from the point of view of gaining information from models using numerical methods of Monte Carlo type stochastic methods are extremely powerful.

In Chapters III and IV we have shown that a Newtonian diffusion process is still well-defined when the density $\rho = |\psi|^2$ has zeros, although the drifts are not defined on the nodes. Indeed, the singularity of the drifts on the nodal surface $N_\rho = \{(x,t)\ \mathbb{R}^d x \mathbb{R}_+ | \rho(x,t)=0\}$

produces a repulsion which is strong enough to keep the configuration from ever reaching the nodal set. Suppose in the stationary case that the nodes of ρ separate $\mathbb{R}^d$ (or more generally the configuration space) into disconnected pieces with no communication among them:

$$\mathbb{R}^d = N_\rho \bigcup_{i=1}^{n} \Gamma_\rho^i \tag{7.5}$$

If the process X_t was started in $X_0 \in \Gamma_\rho^i$ for some i, it will never reach the nodal surface N_ρ and will stay in Γ_ρ^i for all time. The nodal surface N_ρ acts also as impenetrable barrier for the process and X_t is confined in Γ_ρ^i. In conclusion we can say that the family of typical particles is split into several groups by the nodal surface N_ρ of the density and no particle from one group can pass to another.

Remark: If σ is not constant we must consider the Riemannian manifold $\mathbb{M} = (\mathbb{R}^d, g)$, where the metric g is given in terms of the diffusion coefficients σ by $g_{ij} = (\sigma^t\sigma)^{-1}_{ij}$ and a Newtonian diffusion process with values in $\mathbb{M}$ (see Chapter III.). All the conclusions obtained in the case where σ is constant are again valid.

VII.2 Trapping Phenomena and Formation of Spatial Patterns

Impenetrable barriers for diffusions and hence for dynamical systems of quantum theory as well as of stochastic mechanics have been discussed in Chapters III. and IV. These barriers are described by nodal surfaces of the probability density of the underlying diffusion process. Applications to the formation of spatial patterns are manifold and have been given to biological systems [89] as well as several physical dynamical systems [2'] [3,b,c] [4] [13].

VII.2a A Model of the Formation of Jet-Streams in the Protosolar Nebula

It is an old hypothesis that the solar system was formed from a protosolar nebula consisting essentially of a gas of small particles (dust). In one form or another this hypothesis was discussed originally by Descartes (1644), Kant (1755) and Laplace (1796) and has been steadily accompanying all later developments in the discussion of the origin of the solar system. There have been many earlier attempts to explain the origin of the regularity in the distances R_n of the planets from the sun. Classically, this regularity was described by the Titius-Bode law (1766), given R_n in the form

$$R_n = a + bc^n \tag{7.6}$$

for suitable constants a,b,c. One idea which has been intensively discussed recently is a sort of modern version of the Kant-Laplace ring formation: namely that, before the aggregation into planets, concentric roughly planear rings were formed. The rings consist of gas, ice, particles and dust, circulating inside the rings but with no communication with neighbouring rings.

The formation of the planets should then have happened in a later state by aggregation from the jet-streams from Newtonian diffusions. The same kind of ideas can be applied also to the formation of jet-streams around planets (Jupiter, Saturn...).

Our stochastic model provides a general mechanism able of explaining the formation of the jet-streams around a main body (Sun or planet): mutual chaotic collision between dust grains moving in the gravitational field of the central body tend to focus into jet-streams of toroidal shapes centered on the central body. The central body of mass M acts by some spherical symmetric potential $V(|x|)$ and is immersed in some disordered gas of small particles acted upon by V and interacting by collisions or pseudo-collisions, e.g. like in the protosolar nebula of the most common cosmological models. The basic idea consists in thinking of a typical particle in the nebula as performing, under the steady influence of the attraction of the central body and innumerous chaotic collisions with other particles, a stochastic diffusion process. In other words we assume that a typical particle moves along the trajectories of a Newtonian diffusion process X_t with a potential V given approximately by the gravitational attraction and that there exists an invariant distribution as the potential is attractive and the time scale involved is large. Of course the invariant distribution is thought to hold as long as the diffusion approximation is valid.

From the results of Chapter III we then know that the invariant distribution $\rho = |\psi|^2$ is given by the solution of a Schrödinger type eigenvalue problem $H\psi = E\psi$ and that the nodes of ψ act as barriers for the Newtonian diffusion process X_t, hence yielding an explanation for the non-communicating rings in the nebula. The potential being central the eigenfunctions $\psi_{n,l,m}(x)$ in $L^2(\mathbb{R}^3)$ with $l = 0,1,\dots n-1$, $m = -l,\dots,+l$ are of the form

$$\psi_{n,l,m}(x) = R_{n,l}(|x|)\ \psi_l^m(\theta,\varphi) \tag{7.7}$$

with $R_{n,l}$ solution of an ordinary second order differential equation (the radial equation) and $\psi_l^m(\theta,\varphi)$ the usual spherical harmonics. The $|x|$ dependence of the zeros of $\psi_{n,l,m}$ is determined by the zeros of the radial function $R_{n,l}$. Setting $\rho_{n,l,m} = |\psi_{n,l,m}|^2$ we can calculate the associated current velocity $v_{n,l,m}$. The angular momentum in the Z-direction is given by

$$L_Z = \int_{\mathbb{R}^3} dx \, e_Z \cdot (X \times v_{n,l,m}) = c\, m \tag{7.8}$$

with c constant. This is the classical angular momentum of the nebula. Using the conservation of the total classical momentum and choosing Oz along this direction we conclude that the invariant measures to be considered are of the form

$$\rho_{n,l,l}(x) = |\psi_{n,l,l}(x)|^2 \tag{7.9}$$

Recalling now that $\psi_l^l(\theta,\varphi)$ is proportional to $e^{il\varphi}(\sin\theta)^l$ we see that $\rho_{n,l,l}$ is, for l large, concentrated to a small angular region about the equatorial plane. This corresponds to the fact that the planetary system is essentially two-dimensional. The trapping regions ("jet-streams") are regions confined between concentric spheres centered at the center of the main body and two cones. For more details see [3,b], [13] and [4 bis] for numerical results. See also [3c,4] for an application to the morphology of galaxies.

VII.2b Cloud Covering of the Planets

The available picture of planets with a substantial atmosphere exhibits on a large scale regular structures, namely zonal bands. To modelize such phenomena statistical methods are very attractive although it is very hard to justify them from the principles of fluid dynamics.

Indeed think of clouds as being composed of "particles" either droplets of icy flakes. Apart from the gravitational forces, these "particles" feel very complicated forces from the surrounding turbulent atmosphere. We do not intend to take into account the details of these influences but assume that it can be replaced by a diffusion mechanism. Furthermore we shall make no precise statement about the overall force only assuming it is spherical symmetric and derives from a potential $V(r)$. As in the section VII.2.a. the invariant measures to be considered are of the form (7.9). Nodal surfaces are either spheres around the origin corresponding to the zeros of the radial part R_{nl} of the associated wave function or cones defined by the zero of Legendre functions P_l^m. Hence possible zones of confinement are anuli.

This model does not depend too much on different parameters of the planetary atmospheres but accounts for the general feature of these large scale structures [2 , 3c].

The agreement of the model with observations is very good if one keeps in mind that

i) there are few free parameters involved, namely integer n,l,m and it is nice that one can make a fit with relatively low numbers.

ii) the physical parameters in planetary atmosphere vary on a large range as far as the composition, temperature and pressure are concerned, that means that one cannot hope to get a more precise fit to the observations in this model.

VII.2c The Van Allen Radiation Belts

One of the most entertaining results of rockets and satellites investigations has been the discovery in 1958 by Van Allen of zones of radiation which surround the Earth. It was soon established that the source of the observed radiation must be charged particles (i.e. electrons and protons) trapped in the Earth's magnetic field. Assuming that the magnetic field B varies slowly with position then the magnetic moment of a particle is nearly constant. This magnetic moment μ for a particle of mass m is given by

$$\mu = m \frac{v_\perp^2}{2B} \tag{7.10}$$

where $v_\perp$ is the component of the velocity v perpendicular to B From this formula it is clear that the gyrating charged particle will tend to be reflected from regions of higher magnetic fields. Suppose that $v_\parallel$ (the component of v parallel to B) takes the particle into a region of increasing magnetic field. Then $v_\perp$ must increase to keep μ constant but it can increase only until it is equal to the total velocity, at which time $v_\parallel$ has fallen to zero. Thus the particles are reflected back into regions of lower magnetic field. This kind of region where the particles are reflected are called a magnetic mirror. The Van Allen radiation belts are actually of toroidal shape since the particles drift longitudinally owing to the inhomogeneous magnetic field. In a dipole magnetic field electrons drift eastward and protons westward. This suggests that the charged particles are diffused by the irregularities of the electromagnetic field.

The Lorentz force acting on a particle of charge q is given classically by

$$F = q\,v \times B \tag{7.11}$$

v being the velocity. In our stochastic framework we consider (7.11) as the force acting on a charged particle undergoing a Newtonian diffusion with v the current velocity. We do this because the force should be invariant under time reversal. For the stochastic accelaration a of the process X_t we substitute the Lorentz force divided by the mass m. Assuming as explained in Chapter III.6. that the generalized momentum $mv + qA$ is a gradient then $\psi = \rho^{1/2}\, e^{iS}$ satisfies the Schrödinger type equation

$$i\, m\, \sigma^2 \frac{\partial\psi}{\partial t} = \frac{1}{2m}\,(-i\, m\, \sigma^2\nabla - q\, A)^2\psi \tag{7.12}$$

σ being the diffusion constant.

We consider an electromagnetic potential A with axial symmetry producing a magnetic field B in the meridian plane. The most general magnetic field satisfying this requirement can be written as

$$B_x = -x\frac{\partial A}{\partial \rho}\;, \quad B_y = -y\frac{\partial A}{\partial z}\;, \quad B_z = 2A + \rho\frac{\partial A}{\partial \rho} \tag{7.13}$$

A being some function of z and $\rho = (x^2+y^2)^{1/2}$. Introducing now cylindrical coordinates (z, ρ, ψ) (7.12) reduces to a two-dimensional Schrödinger-like equation with an effective potential given by

$$U^{eff}_{\pm|l|} = \frac{1}{2m\rho^2}\,[m\,\sigma^2|l| \pm q\,\rho^2 A(\rho,z)]^2 \tag{7.14}$$

Then if we assume that the vector potential A goes to zero at infinity with $r = (x^2 + y^2 + z^2)^{1/2}$ going to infinity in a given direction the effective potential $U^{eff}_{\pm|l|}$ have the following shape depending on the relative sign of q and l

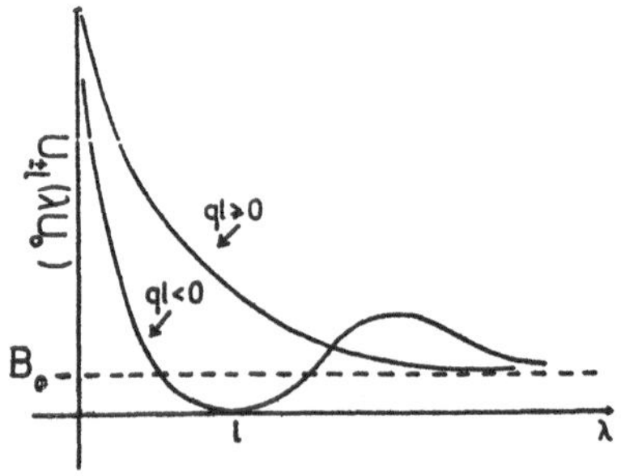

This implies that for $ql \geq 0$ the Schrödinger like equation has no bound state, whereas if $ql < 0$ bound states can appear. To show that under some circumstances the effective potential is confining (i.e. admits bound states) assumptions must be made on the fall-off of the magnetic field at large distances. The effective potential behaves like $q^2\rho^2 A(\rho, z) \simeq B_0\rho^{-1}$. It can be shown that the pure point spectrum of $(-\,i\,m\,\sigma^2\nabla - qA)^2$ is not empty.

The model accounts for the following observational facts:

1) There exists discrete zone of confinements related to the nodal surface of a probability density.
2) In each zone, particles show a general drift either westward or eastward according to the charge, since ql must be positive. Protons drift to the west, electrons to the east.
3) Trapping zones are roughly shaped according to the lines of the magnetic field.
4) Inner belts contain more energetic particles than outer belts.

For a more detailed discussion of the model see [2].

VII.3 A Model for Transport in a Plasma

The basic idea consists of thinking the motion of particles in magnetic structure consisting of numerous small magnetic island as described by the paths of a stochastic non-isotropic differential equation is

$$dX_t = b_t(X_t, t)dt + D\sigma dW_t \tag{7.15}$$

where

$$D = \frac{kT\tau}{m} \tag{7.16}$$

k being the Boltzmann constant, T the absolute temperature, m the electron mass and τ a characterictic term of diffusion inversely proportional to the viscosity. In the slab geometry considered in [53], the magnetic field B_0 acting on the plasma is essentially along the z-axies and the diffusion matrix σ is a symmetric constant matrix of the form

$$\sigma = \begin{bmatrix} \sigma_x & \sigma^* & 0 \\ \sigma^* & \sigma_y & 0 \\ 0 & 0 & \sigma_z \end{bmatrix} \tag{7.17}$$

To determine the drift b_+ we use Newton's law in the mean, which for magnetic force takes the form

$$ma = ev \times B_0$$

a being the stochastic acceleration and v the current velocity of the process X_t. From this dynamical assumption follows that the osmotic velocity u and the current velocity v are solutions of non-linear coupled partial differential equation. It could be shown that these equations can be related to a simpler one of the form (see [52] [53] for more details)

$$\frac{\partial}{\partial t} K(x,y) = \mathcal{D}(x,y)$$

with

$$\mathcal{D} = \frac{1}{2} (\partial j - \frac{ie}{m} Aj) G^{jk}(\gamma_R - \frac{ie}{m} A_R)$$

where A is the vector potential and G the (x,y) part of $g = \sigma\sigma^t$.

Knowing K it is possible to obtain ρ and v and hence the drift b_+. In the simplest case of a stationary process one obtains an Ornstein-Uhlenbeck process namely

$$\begin{matrix} dx_t \\ dy_t \end{matrix} = A \begin{matrix} x_t \\ y_t \end{matrix} + \Delta \begin{matrix} dW_t^{\,x} \\ dW_t^{\,y} \end{matrix}$$

with

$$A = \frac{|e| B_0}{2m} \begin{bmatrix} -\lambda_1/\lambda_2 & -1 \\ 1 & -\lambda_2/\lambda_1 \end{bmatrix}$$

$$\Delta = D \begin{bmatrix} \lambda_1 & 0 \\ 0 & \lambda_2 \end{bmatrix}$$

λ_1 and λ_2 being the eigenvalues of G. One can deduce that the radial diffusion rate satisfies in the case of a strong magnetic field

$$\frac{1}{\tau} \langle x_\tau^2 \rangle = F(\lambda_1, \lambda_2) \frac{kT}{|e| B_0} + o(\frac{1}{\tau})$$

This result corresponds to the simplest situation in which the stochastic differential equation is linear and can be explicitely integrated. The $\frac{1}{B_0}$ behaviour of the radial diffusion rate has been experimentally observed (Bohm's law).

APPENDIX

We review some of the basic notions of probability theory below.

A1. Notations and Conventions

A probability space is a measurable space (Ω, F, P) with $P(\Omega) = 1$. Throughout this book (Ω, F, P) denotes a given complete probability space. This means that (Ω, F) is a measurable space and P is a probability measure on (Ω, F) such that each subset of a P-null set in F is in F. If F_o is a sub-σ-algebra of F the augmentation $\tilde{F_o}$ of F_o is the smallest σ-algebra containing F_o and all the P-null sets in F_o.

Elements of the σ-algebra F are called events. A measurable function $X : (\Omega, F) \to \mathbb{R}^n$ is called a n-dimensional random variable. A funtion X is called measurable if $X^{-1}(A) \in F$ for all Borel sets A in $\mathbb{R}^n$. For any random variable X $\int_\Omega X(\omega)\, P(d\omega)$, if it exists, is called the mean or expectation of X and is denoted by $E[X]$. Thus

$$E[X] = \int_\Omega X(\omega)\, P(d\omega) = \mu$$

$E[X^n]$ is called the nth moment of X about zero and $E[(X-E[X])^n]$ the nth central moment. The second central moment is called variance and will be often denoted by σ^2

$$\sigma^2 = E[(X-E[X])^2] = E[(X-\mu)^2]$$

For any random variable $X : \Omega \to \mathbb{R}^n$ the function $\tilde{P} : \mathbb{R}^n \to \mathbb{C}$ defined by

$$\tilde{P}\,(p) = E[e^{ip\cdot X}]$$

is called the characteristic function of X. Here $p \in \mathbb{R}^n$ and $p\cdot X = \sum_{i=1}^{n} p_i x_i$.

Events $B_1 \ldots B_j$ are called independent if for every $\{i_1 \ldots i_k\} \subset \{1 \ldots j\}$

$$P[\bigcap_{l=1}^{k} B_{i_l}] = \prod_{l=1}^{k} P[B_{i_l}]$$

An arbitrary family of events is called independent if every finite sub-family is independent.

A filtration is a family $\{F_t\}_{t\in I}$ of sub-σ-algebras of F such that $F_s \subset F_t$ for all $s < t$ in I. If the following two conditions are also satisfied then $\{F_t\}_{t\ I}$ is called the standard filtration:

(i) $F_t = F_{t+} = \bigcap_{s>t} F_s$ (right continuity)

(ii) F_o contains all of the P-null sets in F (completeness)

Conditions (i) and (ii) are not obligatory but simplify many technicalities. Indeed many useful theorems for continuous parameter martingales require these hypotheses.

The standard filtration $\{F_t\}_{t\in I}$ associated with a Brownian motion $\{W_t\}_{t\ I}$ is defined by

$$F_t = \sigma\{W_s | o \leq s \leq t\}^{\sim}$$

where the inclusion of the P-null sets in F_t ensure that $F_t = F_{t+}$. The physical meaning of F_t is the following: F_t is the σ-algebra of events occuring up to time t: the "past events up to t".

A2. Conditioning

Let $X : (\Omega,F) \to \mathbb{R}^n$ be a random variable such that $E[X] < +\infty$. Let F_o be a sub-σ-algebra of F. Then there exists a random variable $Y : (\Omega,F_o) \to \mathbb{R}^n$ such that for every A in F_o

$$\int_A Y(\omega)\ P(d\omega) = \int_A X(\omega)\ P(d\omega)$$

If $\bar{Y}$ is any random variable with the same properties then we have $\bar{Y} = Y$ P - almost everywhere.

Y is called the conditional expectation of X with respect (or given) F_o and is denoted by $E[X|F_o]$.

If $F'_o \subset F_o \subset F$ are σ-algebras then

$$E[E[X|F_o]|F'_o] = E[X|F'_o]$$

The following proposition exhibits conditional expectation as a projection on a Hilbert space.

<u>Proposition</u> Let (Ω,F,P) be a probability space and F_o a sub-σ-algebra of F. The $L^2(\Omega,F_o,P)$ is a closed subspace of $L^2(\Omega,F,P)$ and if $\Pi = \Pi^* = \Pi^2$ denotes the orthogonal projector of $L^2(\Omega,F,P)$ on $L^2(\Omega,F_o,P)$ then

$$\Pi f = E[f|F_o]$$

A3. Stochastic Processes

In applications the random evolutions of a system is described by saying that the state of the system is a function of time and randomness. This leads to the following definition: A n-dimensional <u>stochastic process</u> X is a function $X : I \times \Omega \to \mathbb{R}^n$ where I is an interval in $\mathbb{R}^+ = [0,+\infty)$ and $X_t(\cdot) = X(t,\cdot)$ is F-measurable for each $t \in I$. Moreover we say that X has initial value X $\mathbb{R}^n$ if $X_o(\cdot) = x$ a.s. The process X will be denoted by $\{X_t\}_{t\in I}$. The mappings $X(\cdot,\omega):t\to X(t,\omega)$, $\omega\in\Omega$, are called the <u>paths</u> of X.

Given an increasing family $\{F_t\}_{t\in I}$ of σ-algebras on Ω the process is said to be <u>adapted</u> to this family if $X_t \in F_t$ for each $t \in I$ (one says sometimes <u>non-anticipating</u>).

An F-measurable function $\tau:\Omega \to \overline{\mathbb{R}}_+$ is called a <u>stopping time</u> with respect to a filtration $\{F_t\}_{t\in I}$ if and only if $\{\omega|\tau(\omega)\leq t\}\in F_t$ for each $t\in I$. If $\{F_t\}_{t\in I}$ is a standard filtration then the condition on τ is equivalent to $\{\omega|\tau(\omega)<t\}\in F_t$ for each $t\in I$. To any stopping time τ is associated a σ-algebra F_τ, which consists of all sets in $F_\infty = \bigcup_{t\in I} F_t$ satisfying

$$A \cap \{\omega|\tau(\omega) \leq t\} \quad F_t \quad \text{for all} \quad t \in I.$$

If A is any Borel set in $\mathbb{R}^n$ then

$$\tau_A = \inf\{\tau > 0|W_t \notin A\}$$

where W_t is a n-dimensional Brownian motion is a stopping time. In general if τ_1 and τ_2 are stopping times, then $\tau_1 \wedge \tau_2 = \min(\tau_1,\tau_2)$ is a stopping time. In particular if τ is a stopping time then $\tau \wedge t$ is a stopping time for every fixed time t.

One of the basic properties of Brownian motion is the <u>strong Markov property</u>. Loosely speaking this says that given the history of a Brownian motion W up to some finite stopping time τ the behaviour of W after that time depends only on τ and the state W_τ of W at time τ. More precisely if $f: \mathbb{R}^d \to \mathbb{R}$ is a bounded measurable function and τ is a stopping time then

$$E[1\{\omega|\tau(\omega)<+\infty\}\ f(W_{\tau+t})|F_\tau]$$

$$= 1\{\omega|\tau(\omega)<+\infty\}\ E^{W_\tau}[f(W_t)]$$

with $E^{W_\tau}[\cdot] = E[\cdot|W_\tau]$.

A4. Martingales

Let $\{F_t\}_{t\in I}$ be a filtration. A collection $M = \{M_t, F_t\}_{t\in I}$ is called a martingale if and only if

i) $M_t \in L^1$ for each t $(E[|X_t|] < +\infty \ \ \forall t)$

ii) $M_s = E[M_t|F_s]$ for all $s < t$ (almost surely P)

We call M a submartingale if the "=" in (ii) is replaced by $\geq$ and a supermartingale if it is replaced by $\leq$. In other words M is a martingale if and only if both M and $-M$ are supermartingales. The intuitive significance of (ii) is that given the bahaviour of everything up to time s, the value of M at the future time is, on average equal to its value at time s. Thus M_t is the value at time t of our fortune in a game which is fair to us (the game will be unfair for a supermartingale).

A martingale $M = \{M_t, F_t\}_{t\in I}$ is called continuous if and only if

i) $\{F_t\}_{t\in I}$ is a standard filtration

ii) $\{M_t\}_{t\in I}$ has all path continuous

The basic source of continuous martingales is stochastic integrals. A collection $M = \{M_t, F_t\}_{t \ \ I}$ is called a local martingale if and only if

i) M_o is an F_o-measurable random variable

ii) there is a sequence $\{\tau_k\}_{k \ \ \mathbb{N}}$ of stopping times such that $\tau_k \uparrow \infty$ a.s and for each k

$$M^k = \{M_{t\wedge\tau_k} - M_o, F_t\}_{t\in I}$$

is a martingale. The sequence $\{\tau_k\}_{k\in\mathbb{N}}$ will be called a localizing sequence for M.

A stochastic process X_t is said to be a semimartingale if X_t is expressible as a sum of a (local) martingale and a process of bounded variation.

A5. Weak Convergence and Measures on Metric Spaces

It is often useful to consider a stochastic process as a mapping of some measurable space into a space of functions with nice topological properties. Once the space is chosen the stochastic process induces a measure on the function space and the σ-algebra over which the measure is defined is relevant.

Let X be a separable metric space and $\mathcal{B}$ the σ-algebra generated by the open sets. $\mathcal{B}$ is also the smalles σ-algebra with respect to which continuous functions are measurable. The advantage of separability is that the σ-algebra $\mathcal{B}$ is generated by spheres. A measure on X will always be on $(X,\mathcal{B})$. Such a measure μ is always regular in the sense that

$$\mu(A) = \inf_{A \subset G,\ G \text{ open}} \mu(G) = \sup_{C \subset A,\ A \text{ closed}} \mu(C)$$

where A is any Borel set.

Very often it becomes necessary to construct explicitly measures on $(X,\mathcal{B})$ and to this end weak convergence is a useful tool. The problem is to construct a measure μ on $(X,\mathcal{B})$ with certain properties (like having given finite dimensional distribution,...). One constructs a sequence of measures $\{\mu_n\}_{n\in\mathbb{N}}$ having approximate properties and hopefully μ_n will have a limit μ for $n \to +\infty$ which is what we want. Even if the sequence itself does not converge a convergent subsequence will often suffice. This means that we have to define a certain notion of convergence on the space of probability measures on $(X,\mathcal{B})$.

Let M_1 be the collection of probability measures on $(X,\mathcal{B})$. A sequence $\{P_n\}_{n\in\mathbb{N}} \subset M_1$ converges weakly to P if

$$\lim_n \int f \, dP_n = \int f \, dP$$

for each bounded and continuous function f on X. The weak convergence will be denoted by $P = w - \lim_{n\to\infty} P_n$.

This notion of convergence takes advantage of the underlying topology of X, whereas the uniform convergence and the strong convergence do not. It is useful to know that weak convergence arises from a metric defined on $M_1(X)$: indeed there exists a metric d on M_1 such that d-convergence in M_1 is the same as weak convergence. It is also useful to have a condition for the conditional compactness of a set $K \subset M_1$ under weak convergence. Such a condition ensuring that every sequence $\{P_n\}_{n \in \mathbb{N}} \subset K$ has a weakly convergent subsequence exists and is moreover necessary if the metric space X is complete. We give such a condition through the following theorems

Theorem 1

Let X be a compact metric space, then $M_1(X)$ is compact under weak convergence.

Theorem 2

Let X be a metric space and $\{\mu_n\}_{n \in \mathbb{N}}$ be a sequence of measures on $(X,\mathcal{B})$ such that

i) $\mu_n(X)$ is bounded
ii) for any $\varepsilon > 0$ there exists a compact subset K_ε of X such that $\mu_n(X \setminus K_\varepsilon) < \varepsilon$ for all n.

Then the sequence $\{\mu_n\}_{n \in \mathbb{N}}$ has a weakly convergent subsequence.

Theorem 3

Let X be a complete separable metric space. Let $\mathcal{K} \subset M_1(X)$ be a compact subset with respect to the topology of weak convergence. For any $\varepsilon > 0$ there exists a compact subset K_ε such that

$$P(K_\varepsilon) \geqq 1-\varepsilon$$

for all $P \in \mathcal{K}$.

For a proof see K.R. Parthasarathy, Probability measures on metric spaces, Academic Press, New York 1967.

Remark A sequence $\{P_n\}_{n \in \mathbb{N}}$ of probability measures is said to be tight if for each $\varepsilon > 0$ there exists a compact subset K_ε such that $P_n(K_\varepsilon) > 1-\varepsilon$ for all n.

For any probability measure on $\mathbb{R}^k$ there is on some probability space a random variable having that measure as its distribution. Therefore for probability measures satisfying

$$w - \lim_n P_n = P$$

there exist random variables X_n and X having these measures as distributions and satisfying

$$\lim_n X_n = X \qquad \text{(convergence in law)}$$

According to the following fundamental theorem the X_n and X can be constructed on the same probability space and moreover in such a way that

$$\lim_n X_n(\omega) = X(\omega) \qquad \forall\omega$$

a condition which is of course much stronger than the convergence in law.

Skorohod's theorem

Let P_n and P be probability measures on $\mathbb{R}^k$ and $P = \text{w-}\lim_n P_n$ Then there exists random vectors X_n and X on a common probability space $(\Omega,F,\bar{P})$ such that X_n has distribution P_n, X has distribution P and

$$\lim_{n\to\infty} X_n(\omega) = X(\omega) \qquad \forall\omega \in \Omega$$

For a proof see e.g. P. Billingsley, Probability and Measure, John Wiley New York 1979.

A6. Stochastic Ito Integrals

In this section we will briefly discuss the existence of $\int_0^t f(s,\omega)dW_s(\omega)$ where $W_t(\omega)$ is 1-dimensional Brownian motion for a wide class of functions $f : [0,\infty) \times \Omega \to \mathbb{R}$.

Let F_t be the σ-algebra generated by $\{W_s | s \leq t\}$. We call a function of the form

$$f(t,\omega) = \sum_{j\geq 0} e_j(\omega)\, \chi_{[j\cdot 2^{-n},(j+1)2^{-n})}(t)$$

where χ denotes the characteristic function elementary if $e_j(\omega)$ is $F_{j2^{-n}}$ - measurable for all j.

For elementary functions $e(t,\omega)$ we define the integral by

$$\int_s^t e(\tau,\omega)\, dW_\tau(\omega) = \sum_{j\geq 0} e_j(\omega)\,[W_{t_{j+1}} - W_{t_j}](\omega)$$

Now we make the following important observation: if $e(t,\omega)$ is bounded and elementary then

$$E[(\int_s^t e(\tau,\omega)\, dW_\tau(\omega))^2] = E[\int_s^t e(\tau,\omega)^2\, d\tau]$$

From this basic isometry we get an indication of what functions we can extend the integration.

To prove this fundamental relation let $\Delta W_j = W_{t_{j+1}} - W_{t_j}$; then we have

$$\int_s^t e(\tau,\omega)\, dW_\tau = \sum_{j\geq 0} ej(\omega)\, \Delta W_j$$

Since $e_i e_j \Delta W_i$ and ΔW_j are independent for $i<j$ it follows that

$$E[e_i e_j \Delta W_i \Delta W_j] = E[e_j^2]\,(t_{j+1}-t_j)\,\delta_{ij}$$

which implies the basic isometry.

Let S be the class of functions $f(t,\omega) : \mathbb{R}\times\Omega \to \mathbb{R}$ such that

(i) $(t,\omega) \to f(t,\omega)$ is $\mathcal{B}\times\mathcal{F}$-measurable, where $\mathcal{B}$ denotes the Borel σ-algebra on $\mathbb{R}^+$

(ii) For each t the map $\omega \to f(t,\omega)$ is $\mathcal{F}_t$-measurable (i.e. f is adapted)

(iii) $E[\int_s^t f(\tau,\omega)^2\, d\tau] < +\infty$

I(f) $f \in S$ we will define the Ito integral

$$I(f) = \int_s^t f(\tau,\omega)\, dW_\tau$$

I(f) will be $\mathcal{F}$-measurable and

$$E[(I(f))^2] = E[\int_s^t f^2 d\tau]$$

I(f) is a stochastic process called the (stochastic) Ito integral based on Brownian motion.

The idea of the construction is simple: We use the basic isometry to extend in several steps the definition for elementary functions to functions in S. An important property of the Ito integral is that it is a martingale. For continuous martingales we have the following important inequality due to Doob (see e.g.[92 a]):
If M_t is a martingale such that $t \to M_t(\omega)$ is continuous a·s then

$$P[\sup_{0\leq\tau\leq t} |M_\tau| \geq \lambda] \leq \frac{1}{\lambda^p} E[|M_t|^p]$$

provided $E[|M_t|^p] < +\infty$.

Using this inequality and the fact that

$$M_t(\omega) = \int_0^t f(s,\omega)\, dW_s$$

is a martingale with respect to $\mathcal{F}_t$ we conclude that

$$P\left[\sup_{\tau\in[o,t]} |M_\tau| > \lambda\right] \leq \frac{1}{\lambda^2} E\left[\int_o^t f(s,\omega)^2 ds\right]$$

Remark

It is possible to define $\int_o^t f(s,\omega)dW_s$ for a class of functions larger than S.

We finish this section with an example:

$$\int_o^t W_s dW_s = \frac{1}{2}W_t^2 - \frac{1}{2}t$$

The extra term $-\frac{1}{2}t$ shows that the Ito stochastic integral does not behave like ordinary integrals. From this example we see that the image of the Ito integral $W_t = \int_o^t dW_s$ by the map $f(x) = \frac{1}{2}x^2$ is not again an Ito integral but a combination of a dW_s and a ds integral. We have indeed

$$\frac{1}{2}W_t^2 = \int_o^t \frac{1}{2}ds + \int_o^+ W_s\, dW_s$$

It turns out that if we define stochastic integrals as a sum of a dW_s and a ds integral then this family is stable under smooth maps.

A stochastic integral is a stochastic process X_t of the form

$$X_t = X_o + \int_o^t \beta(s,\omega)\, ds + \int_o^t \sigma(s,\omega)\, dW_s$$

The above equation is often written in the shorter differential form

$$dX_t = \beta\, dt + \sigma dW_t$$

Let $g(t,x) \in C^2([o,\infty) \times \mathbb{R}, \mathbb{R})$ then

$$Y_t = f(t,X_t)$$

is again a stochastic integral and

$$dY_t = \frac{\partial f}{\partial t}(t,X_t)dt + \frac{\partial f}{\partial x}(t,X_t)dX_t + \frac{1}{2}\frac{\partial^2 f}{\partial x^2}(t,X_t)\,(dX_t)^2$$

where

$$dt \cdot dt = dt \cdot dW_t = dW_t \cdot dt = 0 \qquad dW_t \cdot dW_t = dt$$

This main result is called the Ito formula, which is very useful for evaluating Ito integrals.
The stochastic integral $\int_o^t \sigma(s,\omega)\, dW_s(\omega)$ based on Brownian motion is due to K. Ito (1941). Stochastic calculus (Ito's formula) based on Brownian motion is carried out according to the rule $(dW_\tau)^2 = dt$.

The theory of stochastic integrals $\int_o^t \Phi(s,\omega)\, dM(s,\omega)$ based on a martingale M is due to Kunita-Watanabe (1967). They also develop a stochastic calculus based on martingale according to the rule $(dM_t)^2 = d\langle M,M\rangle_t$, where M is the so-called quadratic variation of M (see Appendix A 7).

Among spaces of martingales, which may be studied, the space of square integrable martingales is the simplest because of its Hilbert space structure but also the richest to investigate. Indeed the classical types of stochastic integrals discussed in the literature had been introduced as isomorphic transformations of some special space of square integrable martingales.

A7. Definition and Characterization of Quadratic Variation

For $t \in I \subset \mathbb{R}_+$ a partition Π_t of $[0,t]$ is a finite ordered subset $\Pi_t = \{t_0,t_1,t_2,\dots t_k\}$ of $[0,t]$ such that $0 = t_0 < t_1 < \dots < t_k=t$ We denote the mesh of Π_t by

$$\delta\Pi_t \equiv \max_{j=0,1,..k-1} |t_{j+1} - t_j|$$

If $\{\Pi_t^n\}_{n\in\mathbb{N}}$ is a sequence of partition of $[0,t]$, then for each n the members of Π_t^n will be denoted by t_{jn} $j = 0,1,\dots k_n$. The main result is the following theorem.

Theorem

Let $t \in I$ and $\{\Pi_t^n\}_{n\in\mathbb{N}}$ be a sequence of partition of $[0,t]$ such that $\lim_{n\to+\infty} \delta\Pi_t^n = 0$. Suppose M is a continuous local martingale and for each n let

$$\Sigma_t^n = \sum_{t_{jn}\in\Pi_t^n} (M_{t(j+1)n} - M_{tjn})^2$$

Then

i) if M is bounded $\{\Sigma_t^n\}_{n\in\mathbb{N}}$ converges in L^2 to

$$\langle M,M\rangle_t \equiv M^2_t - M^2_o - 2\int_o^t M\, d\, M$$

ii) $\{\Sigma_t^n\}_{n\ \mathbb{N}}$ converges in probability to $\langle M,M\rangle_t$

We call $\langle M,M\rangle_t$ the <u>quadratic variation</u> of M at time t and $\langle M,M\rangle = \{\langle M,M\rangle_t\}_{t\ I}$ the quadratic variation process associated with $\{M_t\}_{t\in I}$.

A process M is a Brownian motion in $\mathbb{R}$ if and only if it is a continuous local martingale with quadratic variation $\langle M,M\rangle_t$ such that

$$\langle M,M\rangle_t = t \quad a\cdot s \quad \text{for all } t.$$

R E F E R E N C E S

[1] ABBOTT L.F., WISE M.B., "Dimension of a Quantum-Mechanical Path", Am. J. Phys. 49 (1981), 37-39.

[2] ALBEVERIO S., BLANCHARD Ph., COMBE Ph., RODRIGUEZ R., SIRUGUE M., SIRUGUE-COLLIN M., "Trapping in Stochastic Mechanics and Applications to Covers of Clouds and Radiation Belts" in "Quantum Probability and Applications II" (ACCARDI L., von WALDENFELS W., Ed., Lecture Notes in Mathematics, 1136 Springer Verlag (1985), 24-39.

[3] ALBEVERIO S., BLANCHARD Ph., HOEGH-KROHN R.,

a. "Feynman Path Integrals and the Trace Formula for Schrödinger Operator", Comm. Math. Phys. 83 (1982), 49-76.

b. "A Stochastic Model for the Orbits of Planets and Satellites: An Interpretation of Titius-Bode Law", Expo. Math. 4 (1983), 365-373.

c. "Diffusions sur une variété riemanienne: Barrières infranchissables et applications", Colloque en l'honneur de L. SCHWARTZ, Astérique 132 (1985), 181-2o1.

d. "Newtonian Diffusions and Planets with a Remark on Non-standard Dirichlet Forms and Polymers" in "Stochastic Analysis and Applications", TRUMAN A., WILLIAMS D. Ed., Lecture Notes in Mathematics 1o95 Springer Verlag (1985), 1 - 24.

e. "Reduction of Non Linear Problems to Schrödinger or Heat Equations: Formation of Kepler Orbits, Singular Solutions of Hydrodynamical Equations" in "Stochastic Aspects of Classical and Quantum Systems", ALBEVERIO S., COMBE Ph., SIRUGUE-COLLIN M. Ed., Lecture Notes in Mathematics 11o9 Springer Verlag (1985), 189-2o6.

[4] ALBEVERIO S., BLANCHARD Ph., HOEGH-KROHN R., MEBKHOUT M., "Strata and Voids in Galactic Structure", submitted to: The Astrophysical Journal.

[4 bis] ALBEVERIO S., BLANCHARD Ph., HOEGH-KROHN R., SCHNEIDER W., "Origin of Hetegonic Jet Streams in the Satellite Systems: A Stochastic Model" BiBoS preprint 1987

[5] ALBEVERIO S., BLANCHARD Ph., GESZTESY F., STREIT L., "Quantum Mechanical Low Energy Scattering in Terms of Diffusion Processes", in "Stochastic Aspects of Classical and Quantum Systems", ALBEVERIO S., COMBE Ph., SIRUGUE-COLLIN M. Ed., Lecture Notes in Mathematics 11o9 Springer Verlag (1985), 2o7-227

[6] ALBEVERIO S., FUKUSHIMA M., KARWOWSKI W., STREIT L., "Capacity and Quantum Mechanical Tunneling", Comm. Math. Phys. 81 (1981), 5o1-513.

[7] ALBEVERIO S., HOEGH-KROHN R.,

a. "A Remark on the Connection Between Stochastic Mechanics and the Heat Equation", J. Math. Phys. 15 (1974), 1745-1747.

b. "Mathematical Theory of Feynman Path Integrals", Lecture Notes in Mathematics, 523 Springer Verlag (1976).

[8] ALBEVERIO S., HOEGH-KROHN R., STREIT L., Energy Form, Hamiltonians and Distorded Brownian Paths", J. Math. Phys. 18 (1977), 9o7-917.

[9] ARNOLD V.I., "Mathematical Methods of Classical Mechanics", Springer Verlag (1978)

[10] BACHELIER L., " Théorie de la Spéculation", Ann. Sci. Ecole Norm. Sup. 17 (1900), 21-86.

[11] BELL J.S.,

a. "On the Einstein-Podolsky-Rosen Paradox", Physics 1 (1964), 195-2oo.

b. "The Theory of Local Beables", CERN preprint TH-2o53 (1975), reproduced in Epistemological Letters (Association Ferd. Gonseth CP 1o81 CH 2o5 Bienne) 9 (1976), 11.

c. "Bertlmann's Socks and the Nature of Reality", Journal de Physique 42 (1981), Suppl. Colloque C2, 41-61.

[12] BILER P., "Stochastic Interpretation of Potential Scattering in Quantum Mechanics", Lett. Math. Phys. 8 (1984), 1 - 6.

[13] BLANCHARD Ph., "Trapping for Newtonian Diffusion Processes", in "Stochastic Methods and Computer Technics in Quantum Dyanamics", Acta Physica Austriaca, Spp. XXVI, (1984), 185-2o9.

[13 bis] BLANCHARD Ph., CARLEN E., DELL'Antonio G.F., "Particle Description in Stochastic Mechanics for Quantum Fields", preprint BiBoS (1987)

[14] BLANCHARD Ph., COMBE Ph., SIRUGUE M., SIRUGUE-COLLIN M.,

a. "Jump Processes in Quantum Theories", in "Stochastic Processes in Classical and Quantum Systems", ALBEVERIO S., CASATI G., MERLINI D. Ed., Lecture Notes in Physics 262 Springer Verlag (1986), 87-1o4.

b. "Path Integral Representation for the Solution of the Dirac Equation in the Presence of an Electromagnetic Field", in "Path Integrals from meV to MeV", GUTZWILLER M.C., INONATA M., KLAUDER S.R., STREIT L. Ed., World Scientific Singapore (1986) 396-413

c. "Stochastic Jump Processes Associated with Dirac Equation" in "Stochastic Processes in Classical and Quantum System", ALBEVERIO S., CASATI G., MERLINI D. Ed., Lecture Notes in Physics 262, Springer Verlag (1986), 65-86.

[15] BLANCHARD Ph., GOLIN S., "Diffusion Processes with Singular Drift Fields", Comm. Math. Phys. 1o9 (1987),421 - 435.

[16] BLANCHARD Ph., GOLIN S., SERVA M., "On Repeated Measurements in Stochastic Mechanics", Phys. Rev. D 34 (1986), 3732-3738.

[17] BLANCHARD Ph., ZHENG W.,

a. "Pathwise Conservation Law for Stationary Diffusion Processes" in "Stochastic Processes in Classical and Quantum Systems", ALBEVERIO S., CASATI G., MERLINI D. Ed., Lecture Notes in Physics 262 Springer Verlag (1986), 1o5-1o8.

b. "Stochastic Variational Principle and Diffusion Processes", in "Stochastic Processes in Classical and Quantum Systems, ALBEVERIO S., CASATI G., MERLINI D., Ed., Lecture Notes in Physics 262 Springer Verlag (1986), 1o9-117.

[18] BORN M., "Zur Quantenmechanik der Stossvorgänge", Z. für Physik 37 (1926), 863-867, [71] ref. therein

[19] BROWN R.

a. "A Brief Account on Microscopical Observations made in the Months of June, July and August, 1827 on the Particles contained in the Pollen of Plants; on the General Existence of Active Molecules in Organic and Inorganic Bodies", Philosophical Magazine N.S. 4 (1828), 161-173.

b. "Additional Remarks on Active Molecule", Philosophical Magazine N.S. 6 (1829), 161-166.

[20] CAMERON R.H., "A Family of Integrals Serving to Connect the Wiener and Feynman Integrals", J. Math. and Phys. 39 (196o), 126-14o.

[21] CAMERON R.H., MARTIN W.T., "Transformations of Wiener Integrals by Non-Linear Transformations", Trans. Amer. Math. Soc. 66 (1949), 252-283.

[22] CARLEN E.,

a. "Conservative Diffusions: A Constructive Approach to Nelson's Stochastic Mechanics", Ph.D Princeton, June 1984.

b. "Conservative Diffusion", Comm. Math. Phys. 94 (1984), 293-315.

c. "Potential Scattering in Stochastic Mechanics", Ann. Inst. H. Poincaré Physics Series 42 (1985), 4o7-418.

d. "Existence and Sample Path Properties of the Diffusions in Nelson's Stochastic Mechanics", in "Stochastic Processes Mathematics and Physics", ALBEVERIO S., BLANCHARD Ph., STREIT L., Ed., Lecture Notes in Math. 1158 Springer Verlag (1986), 25-51

e. "Tail Fields of Some Diffusions with a Limiting Velocity", MIT Preprint

f. "The Stochastic Mechanics of free Scalar Fields", Princeton Preprint (1987).

[23] CARMONA R.,
"Regularity Properties of Schrödinger and Dirichlet Semi-Groups", J. Func. Anal. 33 (1979), 259-296.

[24] CHEBOTAREV A.M., MASLOV V.P.,
"Processus de Sauts et leurs applications dans la Mécanique Quantique", in Feynman Path. Integral; Marseille 1978, ALBEVERIO S., COMBE Ph., HOEGH-KROHN R., RIDEAU G., SIRUGUE-COLLIN M., SIRUGUE H., STORA R. Ed., Lecture Notes in Physics 1o6 Springer-Verlag (1979), 58 - 72.

[25] CHUNG K.L.,

a. "On the Theory of Random Processes", Ann. Math. 41 (194o), 215-23o.

b. "Markov Chains with Stationary Transition Probabilities", second edition, Springer Verlag N.Y. (1967)

[26] CINI M., SERVA M.,
"Stochastic Theory of Emission and Absorption of Quanta", Journal of Physics A19 (1986), 1163-1177.

[27] COMBE Ph., GUERRA F., RODRIGUEZ R., SIRUGUE M., SIRUGUE-Collin M.,
"Quantum Dynamical Time Evolution as Stochastic Flows on Phase Space", Physica 124A (1984), 561-574.

[28] COMBE Ph., HOEGH-KROHN R., RODRIGUEZ R., SIRUGUE M., SIRUGUE-COLLIN M.,

a. "Poisson Processes on Groups and Feynman Path Integrals", Comm. Math. Phys. 77 (198o), 269-298

b. "Feynman Path Integral and Poisson Processes with Classical paths. J. Math. Phys. 23 (198o), 4o5-411

[29] DANKEL, T.G.,

a. "Mechanics on Manifolds and the Incorporation of Spin into Nelson's Stochastic Mechanics", Archive of Rational Mechanics and Analysis, 37 (197o), 192-222.

b. "Higher States in the Stochastic Mechanics of the Bopp-Hagg Spin Model", J. Math. Phys. 18 (1977), 253-255.

[3o] DAVIDSON M.,

a. "A Generalisation of the Fenyes-Nelson Stochastical Model of Quantum Mechanics", Lett. Math. Phys. 3 (1979), 271-277

b. "Momentum in Stochastic Mechanics", Lett. Math. Phys. 5 (1981), 523-529

[31] DE ANGELIS G.F., JONA-LASINIO G.,
"A Stochastic Description of a Spin-1/2 Particle in a Magnetic Field", J. Phys. Math. Gene. A15 (1982), 2o53-2o61.

[32] DE ANGELIS G.F.,
"A Route to Stochastic Mechanics" in "Stochastic Processes in Classical and Quantum System", Ed., Lecture Notes in Physics, 262 Springer Verlag (1986) 16o-169

[33] DE ANGELIS G.F., JONA-LASINIO G., SERVA M., ZANGHI N.,
"Stochastic Mechanics of Dirac Particle in Two Space-Time Dimensions", J. Phys. Math. Gene. A19 (1986) 865-971

[34] DE ANGELIS G.F., JONA-LASINIO G., SIRUGUE M.,
"Probabilistic Solution of Pauli Type Equations", J. Phys. A Math. Gene., 16 (1983) 2433-2444

[35] DE FALCO D., DE MARTINO S., DE SIENA S.,

a. "Position-Momentum Uncertainy Relation in Stochastic Mechanics", Phys. Rev. Lett. 49 (1982), 181-183

b. "Momentum for Sample Paths in Stochastic Mechanics", Lett. Novo Cimento, 36 (1983) 457-46o

[36] DE LA PENA-AUERBACH L., CETTO M.,
"Stronger Form for the Position-Momentum Uncertainty Relation", Phys. Lett., 39A (1972) 65,66

[37] DE WITT-MORETTE C., MAHESHWASI A., NELSON B.,
"Path Integration in Non-Relativistic Quantum Mechanics", Phys. Reports, 56 (1979) 255-372

[38] DIRAC P.A.M.,
"The Principle of Quantum Mechanics", Clarendon Press Oxford, IVth Edition (1958) 125

[39] DOHRN D., GUERRA F.,
"Nelson's Stochastic Mechanics on Riemannian Manifolds", Lett. al Nuovo Cimento, 22 (1978) 121-127

[4o] DOHRN D., GUERRA F., RUGGIERO P.,
"Spinning Particles and Relativistic Particles in the Framework of Nelson's Stochastic Mechanics", in "Feynman Path Integrals, ALBEVERIO S., COMBE Ph., HOEGH-KROHN R., RIDEAU G., SIRUGUE-COLLIN M., SIRUGUE M., STORA R., Ed., Lecture Notes in Physics 1o6 Springer Verlag (1979) 165-181

[41] DOOB J.L.,
"Stochastic Processes", Wiley, New York (1953)

[42] EINSTEIN A.,
"Ueber die von der Molekular-kinetischen Theorie der Wärme geforderte Bewegung von in ruhenden Flüssigkeiten suspendierten Teilchen", Ann. der Physik 17 (19o5) 549-56o.
See also "Investigations on the Theory of Brownian Movement", Edited by R. Fürth, Translated by A.D.Cowper, Ney York, Dutton (1956)

[43] EINSTEIN A., PODOLSKY B., ROSEN N.,
"Can Quantum Mechanical Description of Reality be Considered Complete?", Phys. Rev. 47 (1935) 777-78o

[44] FARIS W.,

a. "Spin Correlation in Stochastic Mechanics", Foundations of Physics, 12 (1982) 1-26

b. "A Stochastic Picture of Spin" in "Stochastic Processes in Quantum Theory and Statistical Physics", ALBEVERIO S., COMBE Ph., SIRUGUE-COLLIN M., Ed., Lecture Notes in Physics 172 Springer Verlag (1982) 154-168

[45] FENYES I.,
"Eine wahrscheinlichkeitstheoretische Begründung und Interpretation der Quantenmechanik", Z. Physik 132 (1952) 81-1o6

[46] FELLER
"An Introduction to Probability Theory and its Applications I and II", John Wiley, New York (1968) 1971

[47] FEYNMAN R.P.,
"Space Time Approach to Non Relativistic Quantum Mechanics", Rev. Mod. Phys. 2o (1948) 367-385

[48] FEYNMAN R.P., HIBBS A.R.,
"Quantum Mechanics and Path Integrals", Mac Graw Hill, (1965)

[49] FREIDLIN M.I., WENTZELL A.D.,
"Random Perturbations of Dynamical Systems", Springer Verlag (1984)

[5o] FUKUSHIMA M.,

a. "Dirichlet Forms and Markov Processes", Kodansha and North-Holland (198o)

b. "Energy Forms and Diffusion Processes", in "Mathematics and Physics", Lecture on Recent Results, vol. 1, L. Streit, World Scientific (1985) 65-97

[51] FÜRTH R.,

a. "Investigations on the Theory of the Brownian Movement", Translated by A.D. Cowper, New Yorky Dotton (1956)

b. "Über einige Beziehungen zwischen klassischer Statistik and Quantenmechanik", Zeitschrift für Phys. 81 (1933) 143-162
See also [71]

[52] GANDOLFO D.,
"I- Equilibre et Perturbation d'un Plasma Torodial avec Flot. II-Modelle de Dynamique Stochastique", Thèse de Doctorat de Troisième Cycle, Marseille (1985

[53] GANDOLFO D., HOEGH-KROHN R., RODRIGUEZ R.,
"Stochastic Model for Plasma Dynamics", to appear in "Processes, Geometry Fields", Proceedings of the 3rd BiBoS Conference, ALBEVERIO S., BLANCHARD Ph., STREIT L., Ed., Lecture Notes in Mathematics, Springer Verlag (1987)

[54] GARBACZEWSKI P.,
"Classical and Quantum Field Theory of Exactly Soluble Nonlinear Systems", World Scientific Publishing Co. (1985)

[55] GIHMAN I.I., SKOHOROD A.V.,

a. "The Theory of Stochastic Processes I, II and III", Springer Verlag (1974) 1975-1979

b. "Stochastic Differential Equations", Springer Verlag (1972)

[56] GIRSANOV I.V.,
"On Transforming a Certain Class of Stochastic Processes by Absolutely Continuous Substitution of Measures", Theor. Prob. App. 5 (196o) 285-3o1

[57] GLIMM J., JAFFE A.,
"Quantum Physics, a Functional Integral Point of View", Springer Verlag, New York (1981)

[58] GOLIN S.,

a. "Uncertainty Relation in Stochastic Mechanics", J. Math. Phys. 26 (1985) 2781-2783

b. "Comment on Momentum in Stochastic Mechanics", J. Math. Phys. 27 (1986) 1549-1555

c. "Indeterminacy Relations in Stochastic Mechanics" in "Stochastic Processes in Classical and Quantum Systems", ALBEVERIO S., CASATI G., MERLINI D., Ed., Lecture Notes in Physics 262 Springer Verlag (1986) 296-3o5

d. "Two Remarks on the Physical Content of Stochastic Mechanics", in "Fundamental Aspects of Quantum Theory", GORIN V., FRIGERIO A., Ed., Plenum (1986) 429-432

[59] GRABERT H. HÄNGGI P., TALKNER P.,
"Is Quantum Mechanics Equivalent to a Classical Process?" Phys. Rev. A19 (1979) 244o-2445

[6o] GUERRA F.

a. "Structural Aspects of Stochastic Mechanics and Stochastic Field Theory", in "New Stochastics Methods in Physics", C. DE WITT-MORETTE, D. ELWORTHY, Ed., Phys. Rep. 77 (1981) 263-312

b. "Probability and Quantum Mechanics. The Conceptual Foundation of Stochastic Mechanics", in "Quantum Probability and Application to the Quantum Theory of Irreversible Processes", L. ACCARDI, A. FRIGERIO and V. GORINI, Eds., Lecture Notes in Math. 1055 Springer Verlag (1984)

c. "Carlen Processes: A New Class of Diffusions with Singular Drifts" in "Quantum Probability and Applications II", ACCARDI L., VON WALDENFELS W., Ed., Lecture Notes in Mathematics 1136 Springer Verlag (1985) 259-267

d. "Quantum Field Theory and Probability Theory. Outlook on New Possible Developments" in "Trends and Developments in the Eighties", ALBEVERIO S., BLANCHARD Ph., Ed., World Scientific Singapore (1985)

[61] GUERRA F., MARRA R.,

a. "Stochastic Mechanics of Spin 1/2 Particle", Phys. Rev. D30 (1984) 2579-2584

b. "Discrete Stochastic Variationnal Principles and Quantum Mechanics", Phys. Rev. D29 (1984) 1647-1655

[62] GUERRA F., MORATO L.,

a. "Momentum - Position Complementarity in Stochastic Mechanics", in "Stochastic Processes in Quantum Theory and Statistical Physics, ALBEVERIO S., COMBE Ph., SIRUGUE-COLLIN M., Ed., Lecture Notes in Physics 173 Springer Verlag (1982) 2o8-215

b. "Quantization of Dynamical Systems and Stochastic Control Theory", Phys. Rev. D27 (1983), 1774-1786

[63] GUERRA L., RUGGIERO P.,
"A New Interpretation of the Euclidean Markov Field in the Framework of Physical Minkowski Space Time", Phys. Rev. Lett. 31 (1973) 1o22-1o25

[64] HEISENBERG W.,
"Über den anschaulichen Inhalt der quantentheoretischen Kinematik und Mechanik", Z. Phys. 43 (1927) 172

[65] HIDA T.,

a. "Canonical Representation of Gaussian Processes and their Applications", Memoire of the College of Science University of Kyoto Series A., vol. XXXIII, Mathematics N° 1 (196o)

b. "Brownian Motion", Springer Verlag (198o), (Applications of Mathematics vol. II)

[66] HIDA T., KUO H.H., POTTHOFF J., STREIT L.,
"White Noise: An Infinite Dimensional Calculus", to appear (1987)

[67] HUDSON R.L., PARTHASARATY K.R.,

a. "Construction of Quantum Diffusion" in "Quantum Probability and Applications to the Quantum Theory of Irreversible Processes", ACCARDI L., FRIGERIO A., GORINI N.V., Ed., Lecture Notes in Mathematics 1o55 Springer Verlag (1984)

b. "Quantum Ito's Formula and Stochastic Evolution", Comm. Math. Phys. 93 (1984) 3o1-323

[68] IKEDA N., WATANABE S.,
"Stochastic Differential Equations and Diffusion Processes", North Holland Publ. (1981)

[69] ITO K.,

a. "Stochastic Integral", Proc. Imp. Acad. Tokyo 2 (1944) 519-524

b. "On Stochastic Integral Equation", Proc. Jap. Acad. 1o4 (1946) 32-35

c. "On Stochastic Differential Equation in a Differentiable Manifold", Nagoya Math. J. 1 (195o) 35-47

d. "On a Formula Concerning Stochastic Differential", Nagoya Math. J. 3 (1951) 55-67

e. "On Stochastic Differential Equations", Memoirės of the Am. Math. Soc. 4 (1951)

f. "The Brownian Motion and Tensor Fields on a Riemannian Manifold", Proc. Int. Congress Math., Stockholm (1962) 536-539

g. "Stochastic Differentials", Applied Math. and Optimization 1 (1976) 374-381

[7o] ITO K.,
"Generalized uniquess complex measures in the Hilbertian metric space with their applications to the Feynman path integral", Proc. of the 5th Berkeley Symposium on Mathematical Statistics and Probability, Vol. II, part 1, University of California Press, Berkeley (1966), 145-161

[71] JAMMER H.,
"The Philosophy on Quantum Mechanics", John Wiley and Sons, New York (1974)

[72] JONA LASINIO G.,
"Stochastic Processes and Quantum Mechanics", Colloque en l'honneur de L. SCHWARTZ, Vol. 2, Astêrisque 132 (1985) 2o3-216

[73] JONA-LASINIO G., MARTINELLI F., SCOPPOLA E.,

a. "New Approach to the Semi-Classical Limit of Quantum Mechanics I", Comm. Math. Phys. 8o (1981) 223-254

b. "The Semi Classical Limit of Quantum Mechanics: A Qualitative Theory via Stochastic Mechanics" Phys. Rept. 77 (1981) 313-327

c. "Tunneling in One Dimension: General Theory, Instabilities, Rules of Calculation, Application", in "Mathematics + Physics", Lecture on Recent Results, Vol. 2, STREIT L., Ed., World Scientific Publishing Co, (1986) 227-26o

[74] JORDAN P.,
"Über eine neue Begründung der Quantenmechanik", Z. Physik 4o (1927) 8o9-838

[75] KAC M.,
"One Some Connections Between Probability Theory and Differential and Integral Equations", Proc. Sec. Berkeley Symp. Math. Stat. and Prob. Berkeley (1951) 185-215

[76] KLAUDER J.,
a. "Continuous Representation and Path Integrals, Revised", in "Path Integrals and their Application in Quantum Statistical and Solid-State Physics", PAPUDOPOULIS G.S., Ed., Plenum, New York (1978) 5-38
b. "Path Integrals, Stationary Phase Approximation and Complex Histories", in "Quantum Fields-Algebras, Processes, STREIT L., Ed., Springer Verlag (198o) 65-9o

[77] KOLMOGOROV A.,
"Fondations of the Theory of Probability", Chelsea Publishing Company, New York (1956)

[78] LANGEVIN E.,
a. "Sur la Théorie du Mouvement Brownien", C.R.A.S. Paris 146 (19o8) 53o-534
b. "Oeuvres Scientifiques de Paul Langevin", Edition du CNRS Paris (195o)

[79] LEVY P.,
a. "Sur les Intégrales dont les Eléments sont des Variables Aléatoires Indépendantes",Ann. Scuola Norm. Pisa 2, N° 3 (1934) 337-366
b. "Sur Certains Processus Stochastiques Homogènes", Comp. Math. 7 (1939) 283-339
c. "Processus Stochastiques et Mouvement Brownien", Hermann, Paris (1965)

[8o] LEWIS S.T.,
"An Elementary Approach to Brownian Motion on Manifolds", in "Stochastic Processes - Mathematics and Physics" ALBEVERIO S., BLANCHARD Ph., STREIT L., Ed., Lecture Notes in Mathematics 1158 (1986) 158-167

[81] LOFREDO M.,
"A Selfconsistent Hydrodynamical Model for the near the Absolute Zero in the Framework of Stochastic Mechanics" Phys. Rev. B35 (1987) 1742-1747

[82] Mc KEAN A.,
a. "Stochastic Integrals", Academic Press, New York (1969)
b. "The Bessel Motion and a Singular Integral Equation", Memoires of the College of Sciences, University of Kyoto, Serie A, Vol. 33, Math. N° 2, 317-322

[83] MADELUNG E.,
"Quantentheorie in hydrodynamischer Form", Z. Physik 40 (1926) 322-326

[84] MASLOV V.P., CHEBOTAREV A.M.,
"Jump Type Processes and their Application to Quantum Mechanics", J. of Soviet Mathematics, 13 (198o) 315-357

[85] MEYER P.A.,

a. "Probabilités et Potentiels", Publications de l'Institut de Mathematique de l'Université de Strasbourg, N° XIV, Paris Hermann (1966)

b. "A Diffential Geometric Formalism for the Ito Calculus"in Stochastic Integrals, Proc. Durham Symp. 198o, WILLIAMS D., Ed., Lecture Notes in Math. 851 Springer Verlag (1981) 256-27o

c. "Séminaire de Probabilité Strasbourg XX (1984-85), Lecture Notes in Math. 12o4 Springer Verlag (1986)

d. "Fock Space and Probability Theory" in "Stochastic Processes Math. and Phys., Bielefeld 1985", ALBEVERIO S ., BLANCHARD Ph., STREIT L., Ed., Lecture Notes in Math. 125o Springer Verlag (1986)

e. "Eléments de Probabilitiés Quantiques", Séminaire de Probabilité Strasbourg XX (1984-85), Lacture Notes in Math. 12o4 Springer Verlag (1986) 186-312

f. "Géométrie Différentielle Stochastique (Bis)", Séminaire de Probabilité XVI, 198o/1981, Supplément: Géometrie Différentielle Stochastique, Lecture Notes in Math. 921 Springer Verlag (1982)

[86] MEYER P.A., YAN J.A.,
"A Propos des Distributions sur l'Espace de Wiener", Séminaire de Probabilité XXI, Lecture Notes in Math. ___ Springer Verlag (1986)

[87] MORATO L.M.,

a. "Path-Wise Stochastic Calculus of Variations with the Classical Action and Quantum Systems", Phys. Rev. D31 (1985) 1982-1987

b. "Path-Wise Calculus of Variations in Stochastic Mechanics", in "Stochastic Processes in Classical and Quantum Systems, ALBEVERIO S., CASATI G, MERLINI D., Ed., Lecture Notes in Physics 262 Springer Verlag (1986) 42o-426

[88] MOYAL
"Quantum Mechanics as a Statistical Theory", Proc. of Cambridge Phil. Soc. 45 (1949) 99-124

[89] NAGASAWA H.,
"Segregation of a Population in an Environment", J. Math. Biology 9 (198o) 213-235

[9o] NELSON E.,

a. "Derivation of the Schrödinger Equation from Newtonian Mechanics", Phys. Rev. 15o (1966) 1o79-1o85

b. "Dynamical Theories of Brownian Motion", Princeton University Press (1967)

c. "Connection Between Brownian Motion and Quantum Mechanics", Mathematical Problem in Theoretical Physics, Proceedings of the VIth International Conference in Mathematical Physics, Berlin (West), 1981, 153 Springer Verlag (1982) 168-179

d. "Quantum Fluctuation - An Introduction in Mathematics and Physics VII", Physica 124A (1984) 5o9-519

e. "Quantum Fluctuations", Princeton University Press (1985)

f. "Field Theory and the Future of Stochastic Mechanics", in "Stochastic Processes in Classical and Quantum System", ALBEVERIO S., CASATI G., MERLINI D., Ed., Lecture Notes in Physics 262 Springer Verlag (1986) 438-469

[91] VON NEUMANN J., "Mathematical Foundation of Quantum Mechanics", translated by BEYER R.T., Princeton University Press, (1955)

[92] PERRIN J.

a. "Mouvement Brownien et Réalité Moléculaire", Ann. de Chim. et de Phys., 8e série, t XVIII (19o9) 5-114

b. "Les Atoms", 4e edition Librarie Aleoun., Paris 1914, (English translation: "Atoms" HAMMICK D.T., van Nostrand N.Y. (1916)

[93] RIESZ F., NAGY B.Sz., "Functional Analysis" 2nd Ed., translated by BORON F.F., with appendix: Extension of Linear Transformations in Hilbert Space which extend beyond this Space by NAGY B.Sz., Federik Ungar Publishing Co. N.Y. (196o)

[94] RÖCKNER H., WIELENS N., "Dirichlet Forms Cosability and Change of Speed Measures", in "Infinite dimensional Analysis and Stochastic Processes, ALBEVERIO S., Ed., Lecture Notes in Mathematics, Pitman, London (1985) 119-144

[95] SCHILPP P.A., "Albert Einstin: Philosopher Scientist", the Library of Living Philosophers, Vol. VII, Evanston, Illinois (1949)

[96] SCHRÖDINGER E.,

a. "Zum Heisenbergschen Unschärfeprinzip", Sitzungsber. Pruss., Akad. Wiss. Phys. Math. Klasse 296 (193o) 296-3o3

b. "Ueber die Umkehrung der Naturgesetze" Berliner Sitzungsberichte (1931) 144-153

c. Sur la Théorie Relativiste de l'Electron et l'Interprêtation de la Mécanique Quantique", Annales de l'Institut H. Poincaré 2 (1932) 289-295

[97] SCHUCKER D.S., "Stochastic Mechanics of Systems with Zero Potential", J. Funct. Anal. 38 (198o) 146

[98] SEILER E., "Stochastic Quantization and Gauge Fixing in Gauge Theories", in "Stochastic Methods and Computer Technic in Quantum Dynamics", Acta Physica Austriaca Spp XXVI (1984) 259-3o8

[99] SERVA M., "Elastic Scattering in Stochastic Mechanics", Lett. Nuovo cumento 41 (1984) 198-2o2

[1oo] SIMON B.,

a. "Functional Integration and Quantum Physics", Pure and Applied Mathematics, A serie of monographs and textbooks, Academic Press (1979)

b. "Instantons, Double Wells and Large Deviations", Publ. AMS Morels (1983)

c. "Quantum Mechanics for Hamiltonian Defined as Quadratic Forms", Princeton University Press (1971)

[1o1] SMOLUCHOWSKI M.V.,

a. "Drei Vorträge über Diffusion, Brownsche Bewegung und Koagulation von Kolloidteilchen", Phys. Zeits. 17 (1916) 557-571

b. "Abhandlungen über die Brownsche Bewegung und verwandte Erscheinungen", Leipzig Akademischer Verlag (1923)

[1o2] STREIT L., HIDA T.,

a. "Generalized Brownian Functionals and the Feynman Integral" in "Stochastic Process and their Application" 16 (1983) 55-69

b. "White Noise Analysis and its Application to Feynman Integral" in "Measure Theory and its application", Lecture Notes in Mathematics", Eds., 1032 (1983)219-226

[1o3] STROOK D.W., "An Introduction to the Theory of Large Deviations", Univ., Springer Verlag (1984)

[1o4] STROOK D.W., VARADHAN S.R.S., "Multidimensional diffusion Process" Springer Verlag (1979)

[1o5] TRUMAN A., LEWIS J.T.,
"The Stochastic Mechanics of the Ground-State of the Hydrogen Atom" in "Stochastic Processes Mathematics and Physics", ALBEVERIO S., BLANCHARD Ph., STREIT L., Ed., Lecture Notes in Mathematics 1158 Springer Verlag Berlin, Heidelberg (1986) 168-179

[1o6] UHLENBECK G.E., ORNSTEIN L.S.
"On the Theory of the Brownian Motion", Phys. Rev. 36 (193o) 823-841

[1o7] VENTZEL A.D.,
"Rough Limit Theorem on Large Deviations for Markov Stochastic Processes"
I. Theory Prob. Applications, 21 (1976) 227-242
II. " " " 21 (1976) 499-512
III. " " " 24 (1979) 675-692
IV. " " " 27 (1982) 215-234

See also [49]

[1o8] WAX N.,
"Selected Paper on Noise and Stochastic Processes", Dover (1954)

[1o9] WERNER R.,
"Screen Observables in Relativistic and Nonrelativistic Quantum Mechanics", Preprint, Osnabrück, FRG (1985)

[11o] WIENER N.,
"Differential Space", J. Math. Phys. 2 (1923) 131-174

[111] WIGNER E.,
"On the Quantum Correction of Thermodynamic Equilibrium", Phys. Rev. 4o (1932) 749-759

[112] YASUE K.,
"Stochastic Calculus of Variations", J. Func. Anal. 41 (1981) 327-34o

[113] YOR M.,
"Filtration de certaines Martingales du Mouvement Brownien", Séminaire de Probabilité XIII, Lecture Notes in Mathematics 721 Springer Verlag (1979) 425-44o

[114] ZAMBRINI J.C.,

a. "Stochastic Dynamics", A review of Stochastic Calculus of Variations, Intern. Journal of Theoretical Physics 24 (1985) 277-327

b. "Stochastic Mechanics According to E. Schrödinger", Phys. Rev. A33 (1986) 1532-1548

[115] ZHENG W.A.,

a. "Seminartingales dans les Variétés et Mécanique Stochastique de Nelson", Thèse Strasbourg, (1984)

b. "Tightness Results for Laws of Diffusion Processes, Application to Stochastic Mechanics", Ann. Inst. H. Poincaré 21 (1985) 1o3-124

[116] ZHENG W., MEYER P.A.,

a. "Quelques Résultats de Mécanique Stochastique", Séminaire de Probabilités XVIII, (1982/83), Lecture Notes in Mathematics 1o59 Springer Verlag (1984) 223-244

b. "Construction de Processus de Nelson Reversible", Osaka J. of Mathematics

c. "Sur la Construction de Certaines Diffusions", Séminaire de Probabilité XX, (1984/85), Lecture Notes in Mathematics 12o4 Springer Verlag (1986) 334-337